RCC、MST 和 NFC 标准及技术应用

王晓华　编著

北京航空航天大学出版社

内 容 简 介

本书重点介绍基于硬件安全技术的移动支付应用系统,涵盖 NFC 技术细节描述和规划、系统安全及密码学、Android 系统中 NFC 的设计和示例代码、13.56 MHz 的硬件和射频技术等;主要内容包括 RCC 限域通信支付系统、MST 磁安全传输支付系统、基于 Android 系统的 NFC 实现框架。

本书适合 RCC、MST 和 NFC 移动支付开发人员阅读。

图书在版编目(CIP)数据

RCC、MST 和 NFC 标准及技术应用 / 王晓华编著. --北京:北京航空航天大学出版社,2018.3
 ISBN 978 - 7 - 5124 - 2676 - 4

Ⅰ.①R… Ⅱ.①王… Ⅲ.①移动通信-通信技术-应用-支付方式-中国 Ⅳ.①F832.2-39

中国版本图书馆 CIP 数据核字(2018)第 048034 号

版权所有,侵权必究。

RCC、MST 和 NFC 标准及技术应用
王晓华　编著
责任编辑　杨　昕

*

北京航空航天大学出版社出版发行

北京市海淀区学院路 37 号(邮编 100191)　http://www.buaapress.com.cn
发行部电话:(010)82317024　传真:(010)82328026
读者信箱:emsbook@buaacm.com.cn　邮购电话:(010)82316936
涿州市新华印刷有限公司印装　各地书店经销

*

开本:710×1 000　1/16　印张:16.75　字数:299 千字
2018 年 4 月第 1 版　2018 年 4 月第 1 次印刷　印数:3 000 册
ISBN 978 - 7 - 5124 - 2676 - 4　定价:49.00 元

若本书有倒页、脱页、缺页等印装质量问题,请与本社发行部联系调换。联系电话:(010)82317024

前言

在当今市面上基于硬件安全技术的移动支付应用系统除了NFC外,还有一些其他的支付应用方案,例如RCC技术和MST技术。但是目前介绍RCC和MST这两项技术的中文书籍较少,业内的朋友曾看过我之前出版的一本关于NFC基础入门的书籍,希望我能把与RCC和MST相关的技术和协议结集成册出版,我没有接受这个建议。

其主要原因有两个:第一,据出版关于NFC技术基础方面书籍的经验,要把答应别人的事情做完,其实不那么容易,即使已经有一些写作材料的积累或者在某一项技术、产品方面有大量的工作经验,但当你准备把它做成一本书时,还是需要付出大量的业余时间和精力去查阅和确认相关资料等,另外要想把它做得好一些、质量高一些,那更是一种压力和责任;第二,当初我正开始准备达到自己下一个阶段工作以外的个人学习目标,即以归零的心态系统地学习一些关于IoT物联网安全和AI人工智能方面的东西,如果答应写书这件事,那么我工作之外的学习重心就可能在一段时间内又会被重新拉回到RCC、MST和NFC的主题上来,这样就需要我在一个近半年的工作以外的时间里做出一个选择。

以前我特别羡慕那些能同时做两件或者多件重要的事情,而且还能把它们做得很好,并且还能来回灵活切换和平衡得不错的人。可惜我天生鲁钝笨拙,尝试过几次同一时间处理两个线程的事,发现每次得到的结果都差强人意,后来也就慢慢想通了、释怀了,觉得把一件事做得好一点这才是最最重要的!如果还能有其他的收获那就需要感激,没有的话也感觉再正常不过。

最后我答应胡编辑来写这本书,自己做出这个决定还是比较迅速的。因为此刻我感兴趣的两件事的时效性还是有本质区别的,其中一件是关于自己制订的那个学习计划,学习本身就是一个长期的过程,不可能是一蹴而就的事。我喜欢的有一个心理仪式感的学习计划的本质是,提醒自己去主动地远离心理上面的舒适区,不要在里面停留太久,空闲时间尽可能地学习和补充一些自己知识结构上面的短板,多接触一些有趣的人和事,总结过往的学习和工作经验,进行资料检查和复核,然后把它们汇总到一本书上去,何尝不是一种学习和有意思的体验呢!

我认为的学习就如罗振宇先生主讲的《罗辑思维》节目中每次上来都反复说的那句话"和你一起终生学习"。当下正处在一个各种技术日新月异、急速变化的时代,不可能说离开学校就可以不学习或者就可以减少知识的汲取。

现在连六七十岁的老人都开始接受新知识、新事物,如智能手机、智能电视的使用等,不学习只能与时代和社会脱节。未来的生活环境会越来越好,那么在接下来的几十年里,依赖传统纸质媒介了解国内外大事和新闻的人将越来越少,更多的是通过网络进行阅读;在有些重要时刻虽然无法亲临现场,但可以使用网络通信工具与亲人们实时互动、分享喜悦。

长辈尚且如此,我辈又怎能不去适应和拥抱那种终生学习的心态呢。我认为上学期间的学习,更多的是去掌握一种适合自己且行之有效的学习方法,并且应多学习和掌握一些基础类学科的知识,如文学、数学、物理和艺术等,其中一些知识有助于在生活和工作中能比较高效地学习和掌握新东西;而大部分知识因为需要比较大块的连续时间的学习才有可能修成一点结果,而这些知识一旦进入社会工作后,又因无法挤出大块时间潜心于此,因此除专业人士或者天才外,对于绝大部分人而言,想在这些领域里自学并有所突破真是非常非常的难!而工作之后的学习呢,我认为有两部分的含义:其一是当下网络经常提及的"刷新认知边界",这个词语是一个比较好的概括,直白点说就是有些知识大量地蕴藏在生活中,在自己身边的人身上,在网络上面,这时就需要自己有一种主动学习的能力,主动地把这些金芝麻给挑选出来为己所用;其二就是新型类的东西,特别是偏应用类的技术,例如新型的商用模式、新型的应用型技术等,学习了这些东西对于工作后的人来讲,就会有比较大的优势。

有人说当下正处于刘慈欣先生的科幻小说《三体》里所解释的"技术爆炸"那样的年代。首先,对于这个观点我个人并不完全认同,因为从近些年诺贝尔科技类的获奖名单和获奖理由看,能够有石破天惊的那种发现并不多。但是近些年来人类在应用技术方面却取得了快速发展,从量子计算、生物科技、医疗技术、无人驾驶、

AR、VR、物联网技术到 AI 人工智能等,确实能明显地感觉应用技术的发展速度越来越快,如果从这个应用技术的发展速度角度去解释"技术爆炸",也确实不为过。

上面提及的这些新型应用技术,量子计算为基于量子力学发展出来的一种应用技术,熟悉《三体》的人都知道里面描述过三体人用来监视地球的"智子"就是靠量子纠缠来运行的。人类从发现半导体技术开始,到当代计算机所遵循的冯·诺依曼架构,现代的信息技术产业如硬件、软件、通信、网络等都是基于这些技术展开的。到目前为止,集成电路芯片的量产制造工艺大多处于 10~14 nm,更小的尺寸工艺如 5~7 nm 正处于研发阶段。尽管科学技术人员还在努力通过各种手段进一步延续晶体管的制造工艺并进一步缩小其尺寸,但是在可预见的未来,将达到控制电子的物理极限。当经典的冯·诺依曼架构无法满足一些超性能架构的计算时,研究新型的计算技术就变得尤为重要,其中量子计算就是当前人类正在攀登的高峰。该技术拥有超强的计算能力,理论数据上显示,拥有 50 个量子比特的量子计算机就能超过当今世界上最先进的超级计算机"天河二号",拥有 300 个量子比特的量子计算机就能将标准服务器需要数万年才能处理完的复杂问题缩短到几秒钟。

我个人认为的"技术爆炸"只是与基础科学相关,与应用科学本质上是没有关系的,那么是不是我们就可以不重视应用科学呢?这个观点我也是不认同的。应用科学有三个重要的特征:第一,针对实际的问题,目的性明确;第二,与人类实践活动的关系密切;第三,直接体现了人类的需求。可以认为应用科学就是我们现代人类生活的空气和水,这能说不重要吗?

本书主要针对的人群是职业技术高中生、在校大学生、移动支付行业的从业者、有工程背景的读者或者自学爱好者们,此书如其书名,是一本介绍硬件安全技术应用于移动支付的书籍。为了让读者看完后能快速地做一些测试实验,书中也穿插了一些示例代码供读者参考。

本书有关 RCC 限域通信技术领域方面的内容得到了泰尔终端实验室高级工程师朱亮博士的大力支持和帮助;全书的框架设计、NFC 技术细节描述和规划、系统安全和密码学、Android 系统中 NFC 的设计和示例代码、13.56 MHz 的硬件和射频技术及新型互联网移动支付市场方面的内容分别得到了杭州雅观科技的田陌晨先生,恩智浦(中国)管理有限公司的陈奕镇先生、罗煜华先生、杨春燕女士、林国辉博士、李竟先生、程恒先生、吴剑先生、徐海先生的鼎力帮助,在此向他们表示衷心地感谢。

最后还要感谢给予大力支持的社群以及小伙伴们:北京航空航天大学出版社、恩智浦中国、小米科技、易宝支付、京东商城、华为北研、Rock NFC、IDEN Team、NFC 移动支付行业交流群、NFC Mobile、NFC Wearable、Mobile CBG、Mobile DD、一起走过的日子、一卡通世界、移动支付网以及许多曾经的同事们,可爱可敬的 Lily 小姐、低调的技术大牛姜彦儒先生、极富洞察力和激情的 Wilson 和 Martin 先生及终身学习者榜样申宇博士等。

由于作者水平有限,书中的错误与不妥之处在所难免,恳请广大读者批评指正。

王晓华

2017/12/6

于北京市海淀区牡丹园

目 录

1 概 述 ·· 1

2 术语和缩略语 ··· 7
  2.1 硬件部分 ·· 7
  2.2 软件部分 ·· 8
  2.3 安全单元和认证部分 ·· 10

3 RCC限域通信支付系统 ·· 13
  3.1 RCC限域通信技术工作原理 ································ 15
  3.2 RCC限域通信协议综述 ····································· 22
    3.2.1 磁通道帧、包、消息数据单元 ························ 23
    3.2.2 射频通道帧、包、消息数据单元 ······················ 26
  3.3 RCC限域通信射频通道帧收发 ····························· 28
  3.4 RCC限域通信技术的物理层特性定义 ····················· 29
    3.4.1 磁通道MC的物理特性 ································ 29
    3.4.2 射频通道RC的物理特性 ······························ 31
  3.5 RCC限域通信技术的会话层命令定义 ······················ 33
    3.5.1 命令交互的频点和地址 ································ 34
    3.5.2 命令消息格式 ··· 36
    3.5.3 INQUIRY激活命令 ···································· 37
    3.5.4 ATI激活响应命令 ····································· 38
    3.5.5 CONNECT_REQ连接请求命令 ···················· 40
    3.5.6 CONNECT_RSP连接响应命令 ···················· 44
    3.5.7 APDATA_REQ数据交换请求命令 ················· 46

    3.5.8　APDATA_RSP 数据交换响应命令 …………………………… 46

    3.5.9　LINKCTL_REQ 维持连接请求命令 …………………………… 47

    3.5.10　LINKCTL_RSP 维持连接响应命令 ………………………… 47

    3.5.11　CHECK1_REQ 冲突检测请求命令 ………………………… 48

    3.5.12　CHECK_RSP 冲突检测响应命令 …………………………… 48

    3.5.13　CHECK2_REQ 连接确认请求命令 ………………………… 49

    3.5.14　LTW 响应方要求等待命令 …………………………………… 49

    3.5.15　CLOSE_REQ 关闭连接请求命令 …………………………… 50

    3.5.16　CLOSE_RSP 关闭连接响应命令 …………………………… 51

 3.6　RCC 限域通信技术与主流近场支付方案的对比 ………………………… 51

4　MST 磁安全传输支付系统 ……………………………………………… 53

 4.1　磁条卡技术原理 …………………………………………………………… 60

    4.1.1　规范协议族 ……………………………………………………… 61

    4.1.2　尺寸参数 ………………………………………………………… 63

    4.1.3　磁道数据详解 …………………………………………………… 66

    4.1.4　磁头原理 ………………………………………………………… 75

    4.1.5　磁道解码应用 …………………………………………………… 79

 4.2　MST 磁传输工作原理 …………………………………………………… 102

 4.3　MST 技术与主流近场支付方案对比 …………………………………… 112

5　基于 Android 系统的 NFC 实现框架 ………………………………… 114

 5.1　基于 Android 系统的 NFC 应用支付框架 …………………………… 202

 5.2　TEE 与 eSE 的应用支付流程示例 ……………………………………… 228

 5.3　卡模拟应用程序示例 …………………………………………………… 235

6　附　录 …………………………………………………………………… 256

1 概述

移动支付技术在全球范围内发展迅速,在中国尤其是以 QR 条码为代表的移动支付应用场景和用户体量更是让人瞠目结舌,人们衣食住行的支付都可以通过一部智能手机来完成,支付场景包括线上与线下。线上支付场景,如在 12306 官网或其他第三方购票平台购买车票时,可以很方便地选择使用支付宝或微信进行应用内支付或扫码支付;线下支付场景,如菜市场、跳蚤市场、夜市小摊、社区小卖部、饭店、宾馆等,现在都支持便捷的手机扫码支付。

上面提到的移动支付技术主要是基于 QR 条码技术。目前这个技术得到大规模的商用主要得益于 3G、4G 网络和智能设备终端的普及,这样就为运维一套闭环的支付体系扫清了底层的技术障碍。对于智能设备终端来讲,只需要支持数据联网以及有基本功能的摄像头和屏幕就可以,并不需要额外的硬件。而对于原先收单的 POS 端硬件系统来讲,只需要增加或者改造一把扫码枪,再简单一点就直接把收单的条码张贴出来,由智能设备终端去对它进行扫描进而完成整个支付流程。这种条码技术的核心以及安全主要放置在后台系统,例如金融支付账户体系、风险控制系统、收单管理系统、清结算网关、绑解卡系统等,都主要放置在后台服务端。可以这样理解,对于前端的条码扫描部分主要完成的是对一个需要进行支付的订单的确认,而鉴权和风险控制管理在扫码部分这里处理的量并不大。

可以看出,现有的基于 QR 条码支付的系统,对终端硬件本身的要求并不高,这从某种意义上预示着两种不同的安全思路:一种是基于底层和硬件技术发展出来的完全被动式的防攻击的设计,它是从底层芯片硬件设计开始做安全防护,到软件安全再到系统安全,这个思路有点类似于"防弹衣"的保护技术,当发现安全隐患或者软硬件系统被攻破时,进行安全升级的主要是通过事后打补丁或再多穿几件"防弹衣"实现的,属于一种完全被动式的安全防守方式;另外一种是以互联网企业为主发展出来的主动防守的风险控制管理方式,通过结合已有的大量用户互联的行为数据,包括支付场景的时间、地点、行为、习惯、逻辑等,当发现有异常支付行为时会主动提升安全等级的验证要求,如果评定为恶性攻击或者支付环境不满足,那么系统就可以终止这笔交易，这样相当于把风险提前识别出来，达到一个主动防守的功效。

目前，主流市场对于上面两种安全思路的看法：未来它们之间不会是谁代替谁的问题，而是如何把二者融合得更好更紧密的问题，既能把两种安全方式的优势利用起来，又能给消费者带来极致的用户体验，这是未来一个很重要的课题。包括 IoT 物联网的安全，未来海量的设备终端会连接到主干网上，如果这些智能设备不能保证连接是安全的，想象一下当汽车在启动无人驾驶时，控制系统被攻破和被坏人接管了，或者晚上休息后，家中厨房的那些智能设备生火做饭了，城市公交指挥系统如红绿灯和网络摄像头等完全失去了控制，那后果是相当可怕的。还有一个就是现在发展得如火如荼的区块链技术，这个技术从本质上讲就属于上面提到的第二种主动防守技术，它是这类技术的代表，假如这个技术应用在当下的移动支付场景中，就相当于对每一次的交易数据节点都会做周边节点的安全确认，只有当别的节点也能证明这笔交易是安全可信的时，此次交易才会发生。例如，张三突然发起一笔面对面 QR 扫码，支付购车款项到李四那边，那么当系统认为这笔款项属于额度较大的款项，且对此笔交易的时间、地点和用户行为有疑虑时，会开启例如手机短信验证码或人脸识别确认，待短信验证码通过或人脸识别没有问题后，后台系统才会对李四是否真实拥有这辆车的相关节点的数据进行确认，当发现多于 3 个节点的数据都无法证明李四拥有这笔财产时，会标识这笔交易风险等级较高，如需要延时 24 小时或 48 小时后才能进行款项交接，这样就可以给张三足够的时间去检查和确认这笔交易。

综合上面说到的 IoT 物理网和基于区块链技术支付的例子，其涉及两个问题：第一，对于这种海量的智能设备连接到网络后，如果全部的安全由云端来保护，不管对于技术实现还是负荷来讲都是天方夜谭，想象一下如果智能设备端没有安全保护措施，那攻击方是完全可以轻易仿照出伪终端来欺骗服务端并且发起对云端的攻击的，所以在设备终端重点采用"防弹衣"式的被动防守模式，可以确保需要接入的设备终端是安全可信的，不能被仿制的；第二，即使使用了类似区块链的技术，核心问题还是如何保证遍历的周边节点的数据是安全可靠的，前提是不仅需要所有的数据已经备份到了云端，而且是数据在备份上传时是安全可靠的，这样就必须结合终端的"防弹衣"式的被动安全防守，再加上云端大数据处理后的主动安全防卫。

本书重点描述的是基于硬件安全技术的移动支付应用系统，对于云端主动安全防卫的技术不做过多介绍。那么在现有的基于硬件安全方式的应用系统主要是以

1 概 述

NFC 技术结合 SE 安全单元为主的,例如目前苹果公司的 Apple Pay、华为公司的 Huawei Pay 和小米公司的 Mi Pay 都是以 NFC 技术为主,可以支持手机银行卡或者城市公交卡等业务,另外三星公司的 Samsung Pay 底层技术既包含 NFC 技术也包含 MST。本书重点介绍的是 NFC 结合 SE 安全单元的应用技术,虽然 MST 技术也可以理解成是一种基于硬件的支付应用系统,但是 MST 主要还是一项传输技术,本身对于硬件安全并没有过多的涉及,单独拉出来看这项技术,它和 NFC 传输技术比较像,都是一个在支付场景中实现的一个数据传输功能,本身并不会涉及过多的安全。

在国内以国民技术公司为主推广的 2.45 GHz RCC 技术,从技术创新层面来讲可圈可点的地方确实很多,特别是在终端支付设备上用户无需更换或者改造手机,只需要更换一张 SIM 卡就可以,非接触的射频前端、天线以及原来的 SIM 卡的电信功能全部集成在一张 SIM 卡上,天线就内嵌在 SIM 卡上。RCC 技术最大的优势在于基于 2.45 GHz 的通信频率还能控制通信和交易的距离,无需特别硬件的手机,只要支持 SIM 卡接口的智能设备就可以;最大的缺点就是,需要去改造或者增加已经在市面上运营的 13.56 MHz 的 POS 机或者闸机读头以实现对 RCC 的支持,另外就是 RCC 技术整个供应链上的厂家没有支持 13.56 MHz 的 NFC 的公司多,所以要去推动重新改造一番底层的基础支付设备终端确实非常有挑战性。综上也就解释了为什么国内外现在主流的银行或者公交的非接触支付系统使用 13.56 MHz 频率更多的原因,其他的非接触频段支付主要应用在一些相对闭环的场景。

相较于以扫码为代表的互联网移动支付技术,以 NFC 和 RCC 为代表的近场支付技术具有多项优势:第一,如上面介绍过的,除去服务器端提供风险安全控制外,支付终端采用硬件安全机制,支付环境的安全优势比条码支付的要显著许多;第二,可以支持离线支付,例如在无网络或者网络信号差的地铁里,支付场景不受网络环境影响,可以真正实现全场景的支付覆盖;第三,对于一些要求高稳定、高频次、高密度的支付场景,例如公交车的振动、上下班地铁公交系统的高人次通行、高亮度阳光对屏幕的反射等问题,对于扫码支付来讲就会是比较大的问题,而对于射频非接触的支付方式就不会有类似的问题。

NFC 技术是从 13.56 MHz 的 RFID 基础上发展起来的一项技术,从单独的 NFC 技术本身来讲已经发展了十几年,所以在成熟的商业应用方面对比其他的射频段或者非接触技术(例如 RCC 技术)就会有很大的继承关系的优势,例如在公交领域、芯片银行卡应用和一些行业应用方面,它们大部分都使用13.56 MHz这个通信频段,而且在数据通信链路协议上面,NFC技术采用了与ISO/IEC 7816接触技术相同的应用数据协议。这样对于支付系统来讲,POS机或者读头可以不需要改动,重点改造智能支付终端就可以,例如在智能手机中需要加入NFC和SE的硬件模块,如果希望NFC支持READER和P2P的功能则需要集成NFC协议栈来支持。

从2002年3月25日索尼公司与荷兰飞利浦公司(后来独立拆分出半导体业务成立了恩智浦半导体公司)共同宣布了一个NFC的粗略技术框架开始,再到后来加入的诺基亚公司,经过多方长时间的磨合,包括协商出来P2P点对点传输技术,在底层射频通信层面同时适配支持了恩智浦半导体为主导的Type A技术和索尼公司主导的Type F技术,并且在诺基亚功能机时代就尝试加入NFC的支付应用,但是NFC技术一直处于不温不火的境况。直到2010年Google公司发布了Nexus S并且把NFC的协议栈开源出来,相当于把NFC正式拉上了高速列车,这个时候NFC技术才在运营商、银行、公交公司得到了蓬勃发展。但是多方商业机构因为支付的主导权问题,大家又进入了一个群雄争霸的时代,SE安全模块是选择放置在SIM卡里,由运营商来主导;还是SE安全模块放置在扩展SWP接口的SD卡上,由银行机构来管理;还是SE内嵌在手机电路板中,由OEM厂家来主导。不过庆幸的是争执的这几年技术本身并没有停滞下来,全球范围内以中国的三大运营商为主力军,在持续不断地给NFC产业输入血液,运营商主导的SIM卡方案并没有取得消费者的完全认可。

再接下来就是2016年,当Apple Pay在中国市场上推出时确实形成了不小的化学效应,其分别推出了线下非接触支付和集成在应用程序端或网页端的支付应用。前者使用的场景为可以通过网络新发的一张卡或者绑定一张银行卡到手机里,然后把手机靠近POS机就可以完成支付;后者则可以在第三方开发的应用程序或者网站上面集成Apple Pay提供的SDK套件,这样就可以通过手机硬件上面提供的SE安全单元里的支付程序完成支付。这些功能在苹果公司最新的官方产品定义中把它们分别叫做Pay in Stores、Pay within Apps和Pay within Websites,后两者其实就相当于把线上和线下打通了,在线上选购完商品后,通过线下手机内置的SE安全模块提供的安全支付通道支付。对比上面刚讲的QR条码支付体系几乎完全依靠云端保证安全,如能把这种已经内嵌SE安全模块的功能结合使用,那么会有一个很大的安全提升和用户体验提升。

1 概 述

除去上面介绍的一些近距离通信技术以外,平时生活中接触得比较多的无线电连接技术有用途广泛的移动电话、Wi-Fi 和音响等,在把通信的距离缩小到半径 200 m 左右及通信支持上下行技术的情况下,我们能看到下面将要介绍的这些已有的技术。但是当下使用这些技术进行支付或者移动支付的应用并没有或者准确来说并不是行业主流,毕竟技术设计的方向不尽相同,所以本书也不会过多地介绍下面几种技术的细节。

蓝牙无线技术:当初设计这个技术的时候就是为了代替两台手机或电脑之间传送数据时使用的线缆,现在它们之间的理论通信距离半径为 10 m 左右,目前广泛应用于蓝牙耳机、音响以及车载电子设备。

Wi-Fi 技术:在一个相对比较固定的环境下使用这个技术能大大降低室内布线的复杂度,例如在宾馆、家庭、办公室等这个技术随处可见。这个技术设计的初心是为了能更好地优化局域网络(LAN),在一个相对没有金属障碍的环境里使通信距离能到达半径 100 m 或更远。

ZigBee 无线技术:主要应用于组网规模相对较大的一些工业自动化领域,这个技术的通信范围在半径 100 m 内,目前广泛应用于餐馆使用的手持订单终端。

红外无线技术(IrDA):是一个短程小于 1 m 的通过红外线进行数据交换的技术,红外线接口经常用于电脑、手机和数码产品之间。

无线射频识别(RFID):此技术本身的定义比较宽泛,有高低频之分,以及有源和无源通信之分。其工作原理为通过读头可以发起对外部的无线识别标签进行远程存储和检索数据。

非接触技术:一般大家说的这个技术都是约定俗成的泛指 ISO 14443 和日本的 FeliCa 技术。这个技术其实也是属于 RFID 中的一种,主机读头为有源工作,而卡片则是无源的。在工作时读头发起射频场强并且附加通信数据给卡片,卡片在收到无线射频场强时会把这个能量转换成工作电压供自己启动和工作,并且能快速处理发送过来的数据信息进行响应和处理。

NFC 与其他短距离通信协议的比较如表 1-1 所列。

综上所述,本书基于硬件安全技术的移动支付应用系统会把重点放在 MST、2.45 GHz RCC 和 NFC 技术基于 Android 系统的应用方案介绍。

表 1-1 NFC 与其他短距离通信协议的比较

通信参数	NFC	Bluetooth	Bluetooth Low Energy	RFID	IrDA
无线射频兼容	ISO 18000-3	主动模式	主动模式	ISO 18000-3	主动模式
参考标准	ISO/IEC ECMA ETSI NFC Forum	Bluetooth SIG	Bluetooth SIG	ISO/IEC	ISO
网络协议	ISO 13157 etc.	IEEE 802.15.1	IEEE 802.15.1	ISO 13157 etc.	
网络类型	P2P	WPAN	WPAN	P2P	P2P
密码技术	无	有	有	无	无
通信距离	约 10 cm	约 10 m(class 2)	约 100 m	约 10 cm	约 1 m
通信频率	13.56 MHz	2.4~2.5 GHz	2.4~2.5 GHz	13.56 MHz	3.8 MHz
通信速率	424 kbit/s	2.1 Mbit/s	约 1.0 Mbit/s	424 kbit/s	115 200 kbit/s（SIR）
准备时间	<0.1 s	<6 s	<0.006 s	<0.1 s	<0.5 s
功耗	<15 mA(读/写)		<15 mA 接收/发送	约 250 mA	约 300 mA
易用性	方便、快捷	一般	一般	快捷	方便
交互性	高效、安全	要求配对	要求配对	高效	无障碍
应用场景	支付、门禁、名片分享、配对	大数据交换	大数据交换	物流追踪	控制与数据交换

2 术语和缩略语

本书中将会出现一些专业术语等,其中大量的是以英文字母缩写为主的。本章主要介绍书中出现或者使用的符号、术语和缩略语,阅读本章有助于读者对术语的形成有一个大概的了解,以方便阅读本书。

2.1 硬件部分

硬件类术语和缩略语如表2-1所列。

表2-1 硬件类术语和缩略语

标 识	出处和解释
RFID	Radio Frequency Identification,无线射频识别技术
NFC	Near Field Communication,近场通信技术
RCC	Range Controlled Communication,限域通信
SE	Secure Element,安全单元
eSE	embedded Secure Element,嵌入式安全单元
CLF	Contact Less Front-end,射频前端
CLT	Contact Less Tunnel,接触通道
SiP	System in Package,系统级封装
SMT	Surface-Mount Technology,表面贴装技术
BB	BaseBand,基带
AP	Application Processor,应用处理器
IoT	Internel of Things,物联网
AI	Artificial Intelligence,人工智能
VR	Virtual Reality,虚拟现实
AR	Augmented Reality,增强现实
PCB	Printed Circuit Board,印刷电路板
PCBA	Printed Circuit Board Assembly,印刷电路板组件

续表 2-1

标　识	出处和解释
ECO	Engineering Change Orde,工程更改号
I^{2}C	Inter-Integrated Circuit,I^{2}C 总线
ACK	Acknowledgment,应答响应
UART	Universal Asynchronous Receiver Transmitter,通用异步收发传输器
SPI	Serial Peripheral Interface,串行外设接口
SIM	Subscriber Identification Module,用户身份识别卡
UICC	Universal Integrated Circuit Card,通用集成电路卡片
SWP	Single Wire Protocol,单线协议
SWIO	Single Wire protocol Input/Output (aka SWP),单线协议包括信号进和出
CICC	Close-Coupled Integrated Circuit Card,密耦合卡片
CCD	Close-Coupled Device,密耦合读头
PICC	Proximity Integrated Circuit Card,近耦合卡片
PCD	Proximity Coupling Device,近耦合读头
VICC	Vicinity Integrated Circuit Cards,疏耦合卡片
VCD	Vicinity Coupling Device,疏耦合读头
UID	Unique Identifier,唯一标识号码
POS	Point of Sale,销售终端
MC	Magnetic Channel,磁通道
RC	RF Channel,射频通道
DME	Differential Manchester Encoding,差分曼彻斯特编码
ISM	Industrial Scientific Medical,工业、科学和医用频段
MST	Magnetic Secure Transmission,磁安全传输

2.2　软件部分

软件类术语和缩略语如表 2-2 所列。

表 2-2　软件类术语和缩略语

标　识	出处和解释
SDK	Software Development Kit,软件开发工具包

2 术语和缩略语

续表 2-2

标 识	出处和解释
NDK	Native Development Kit,原生开发工具包
JVM	Java Virtual Machine Java,虚拟机
JSR	Java Specification Requests,Java 规范申请
JNI	Java Native Interface,Java 原生接口
FRI	Forum Reference Implementation,论坛参考实现
NDEF	NFC Data Exchange Format,NFC 数据交换格式
RTD	Record Type Definition,记录类型定义
HAL	Hardware Abstraction Layer,硬件抽象层
DAL	Data Access Layer,数据访问层
OSAL	Operating System Abstraction Layer,操作系统抽象层
HCI	Host Controller Interface,主机控制接口
LLC	Logical Link Control,逻辑链路控制
LlCP	Logical Link Control Protocol,逻辑链路控制协议
P2P	Peer to Peer,点对点
REQA,B	Request Command, Type A,Type B,请求命令类型 A 卡,请求命令类型 B 卡
ATQA,B	Answer to Request, Type A,Type B,应答请求类型 A 卡,应答请求类型 B 卡
SAK	Select AcKnowledge,选择应答
APDU	Application Protocol Data Unit,智能卡应用协议数据单元
C-APDU	Command APDU,命令型 APDU 数据
R-APDU	Response APDU,应答型 APDU 数据
ATR	Answer to Reset,复位应答命令
ATTRIB	PICC Selection Command, Type B,类型 B 卡选择命令
HLTA,B	Halt Command, Type A,Type B,停止命令类型 A 卡,停止命令类型 B 卡
SEL	Select Code, Type A,类型 A 卡选择代码
SELECT	Select Command, Type A,类型 A 卡选择命令
WUPA,B	Wake-UP Command, Type A,Type B,复位命令类型 A 卡,复位命令类型 B 卡
NfcA	ISO14443-3A,ISO14443 第三层协议类型 A 卡
NfcB	ISO14443-3B,ISO14443 第三层协议类型 B 卡
NfcF	JIS 6319-4,日本 Felica 标准的参考规范
NfcV	ISO 15693,近场耦合卡片的参考规范

续表 2-2

标　识	出处和解释
IsoDep	ISO 14443-4 近场感应卡片的参考规范
CLT cmd	First byte is not 'E0(RATS)', '50(HLTA)', '93(ANTI-1)', '95' or '97', 第一条指令不为 E0/50/93 的命令
Tag 1	Topaz 企业内部商标卡片的叫法（T1T）
Tag 2	MIFARE Ultralight 企业内部商标卡片的叫法（T2T）
Tag 3	Felica 企业内部商标卡片的叫法（T3T）
Tag 4	MIFARE Desfire 或者 ST25TA 系列企业内部商标卡片的叫法（T4T）
Tag 5	I·CODE 或者 ST25TV 系列企业内部商标卡片的叫法（T5T）

2.3　安全单元和认证部分

安全单元和认证的相关术语和缩略语如表 2-3 所列。

表 2-3　安全单元和认证的相关术语和缩略语

标　识	出处和解释
GCF	Global Certification Forum，全球认证论坛
PTCRB	PCS Type Certification Review Board，个人电脑类型认证审查委员会
CE	Conformite Europeenne，欧洲安全合格标识
FCC	Federal Communication Commission，美国联邦通信委员会
FCT	Forum Certificate Tests，论坛认证测试
PBOCx.x	People's Bank of China (aka JR/T 0025—2005)，中国人民银行支付规范标准
OSCCA	Office of Security Commercial Code Administration，中国国家密码管理局
BCTC	Bank Card Test Center (China)，银行卡检测中心（中国）
SCC	Standards Council of Canada (Canada)，加拿大标准委员会（加拿大）
CESTI	Centres d'Evaluation de la Sécurité des Technologies de l'Information (France)，技术安全信息评价中心（法国）
BSI	Bundesamt für Sicherheit in der Informationstechnik (Germany)，德国联邦信息安全局（德国）
UKAS	United Kingdom Accreditation Service (UK)，英国认证服务局（英国）
NIST	National Institute of Standards and Technology (US)，国家标准与技术研究所（美国）

2 术语和缩略语

续表 2-3

标识	出处和解释
CC	Common Criteria,通用标准
EAL	Evaluation Assurance Level,评估保证级别
FIPS	Federal Information Processing Standards,联邦信息处理标准
EMV	Europay MasterCard and VISA,欧洲万事达和维萨组织
JCOP	Java Card Open Platform,Java 卡开放平台
JCVM	Java Card Virtual Machine,Java 卡虚拟机
JCRE	Java Card Runtime Environment,Java 卡运行环境
GP	Global Platform,全球卡片平台标准
OP3	Open Platform Protection Profile,开放平台保护规范
TOE	Target of Evaluation,目标评价
SFRs	Security Functional Requirements,安全功能要求
SARs	Security Assurance Requirements,安全保证要求
PP	Protection Profile,保护规范
ST	Security Target,安全目标
DAP	Data Authentication Pattern,数据认证模式
ACM	Access Condition Matrix,访问状态矩阵
EAC	Extended Access Control,扩展访问控制
LDS	Logical Data Structure,逻辑数据结构
CQM	Card Quality Management,卡片质量管理
PKI	Public Key Infrastructure,公共密钥
RNG	Random Number Generator,随机数发生器
SFI	Single Fault Injection,单一故障注入
AES	Advanced Encryption Standard,高级加密标准
DES	Data Encryption Standard,数据加密标准
RSA	Rivest, Shamir and Adleman asymmetric algorithm,不对称算法
IIN	Issuer Identification Number,发行识别号码
PIN	Personal Identification Number,个人识别号码
OID	Object Identifier,对象标识符
TLV	Tag Length Value、标签、长度、键值
SMX	Smart MX,企业内部商标卡片的叫法
CM	Card Manager,卡片管理者
CVM	Cardholder Verification Methods,持卡人检验方法

11

续表 2-3

标识	出处和解释
ISD	Issuer Security Domain,主控安全域
SSD	Supplementary Security Domain,辅助安全域
TSM	Trusted Service Management,可信服务管理
HSM	Hardware Security Module,硬件加密机
SEI-TSM	Secure Element Issuer-Trusted Service Management,安全单元发行商可信服务管理
SP-TSM	Service Provider-Trusted Service Management,服务提供商可信服务管理
TEE	Trusted Execution Environment(ex. MobiCore),可信赖执行环境
OP-TEE	Open Source Trust Execution Environment(https://github.com/OP-TEE),TEE 开源社区操作系统
REE	Rich Execution Environment(ex. Android),通用执行环境
TA	Trusted Application(ex. Alipay Wallet),可信赖应用程序
TZ	TrustZone(ex. Cortex-A15,Cortex-A9,Cortex-A8,Cortex-A7,Cortex-A5,ARM1176),信任区
TCM	Tightly Coupled Memory,紧耦合存储器
TZPC	TrustZone Protection Controller,信任区保护控制器
TZMA	TrustZone Memory Adapter,信任区存储适配器
TZASC	TrustZone Address Space Controller,信任区控制器地址空间
SMC	Secure Monitor Call,安全监控电话
SCR	Secure Configuration Register,安全配置登记
SE-Linux	Security Enhanced Linux,安全增强式 Linux
QRCode	Quick Response Code,二维码
DES	Data Encryption Standard,数据加密标准
3DES	Triple Data Encryption Standard,三重数据加密标准
RSA	Ron Rivest,Adi Shamir,Leonard Adleman,RSA 加密算法
AES	Advanced Encryption Standard,高级加密标准
SM2	椭圆曲线公钥密码算法
SM3	密码杂凑算法
SM4	分组密码算法(原来也叫 SMS4)
CBC	Cipher Block Chaining,密文块链接模式
ECB	Electronic Code Book,电子密码本模式
PAN	Primary Account Number,主账号
TSP	Token Service Provider,标记服务提供商

3 RCC 限域通信支付系统

RCC(Range Controlled Communication)限域通信,基于 2.45 GHz 频段并结合低频磁技术,主要原理为移动支付方案全部集成到一张 SIM 卡或者 SD 卡中,对于支付终端本身无需做额外的硬件改动,只需要更换支持 RCC 的 SIM 卡或者 SD 卡就可以支持小额支付,应用场景如手机、银行卡、公交卡、商户消费卡、停车卡、加油卡、校园一卡通、VIP 会员卡、图书借阅卡、优惠券、电子票等。

RCC 产品形态比较简单,主要分为卡端和读卡器端两类。第一类 RCC 卡端,例如市面上支持 RCC 功能的 SIM 卡和 SD 卡等,目前支持 RCC 的 SIM 卡类型也已经涵盖了标准 SIM 卡、MicroSIM 卡和 NanoSIM 卡。众所周知射频通信在接收区域内如有金属物体的话,那么金属物体对该频段的射频会产生折射和反射从而影响射频接收器信号的读/写,但是通过射频参数调试和适配过的 RCC SIM 卡则可以兼容市面上金属外壳的手机。第二类 RCC 读卡器端,为配合读取 RCC 卡片,受理终端也需要支持 RCC 功能。根据不同的需求场景可采用不同的 POS 机读头改造方案。方案一,对于研发阶段或有升级需求的终端机具,可以通过在机具电路板中集成 RCC 读卡器芯片或读卡器模块的方式实现。RCC 读卡器模块具有多种尺寸规格,采用例如 UART、I^2C 或者 SPI 串行等接口,使用简洁的指令集,非常利于终端机具集成。方案二,对于已经投放到市场的 13.56 MHz 频段的终端机具,可以采用 RCC 读头转接模块的方式实现,也就是贴膜方式。其主要原理是通过转接天线将终端发出的 13.56 MHz 信号转换成 RCC 卡片的 2.45 GHz 信号,从而实现终端机具与 RCC 卡片之间的数据交互。此方式无需改动终端机具中的嵌入式软件,仅需从机具上取电,将转接天线叠放在 13.56 MHz 的天线上即可,实施起来简单快捷。RCC 读头转接模块天线的尺寸和材质需要根据不同型号的机具进行定制和调试。

如图 3-1 所示为 RCC 限域通信框图。RCC 限域通信的基本原理是:在通信开始前,RCC 读卡器通过 2 kHz 的低频磁通道发送接下来用于射频通道通信的参数和密钥,并结合发送方和接收方的磁感应原理,以确保其两端之间的射频通信距离能控制在 10 cm 以内。RCC SIM 卡则在接收到发送端通过低频磁通道传过来的参数和密钥后去激活 2.45 GHz 的射频通道,然后在 RCC 读卡器与 RCC SIM 卡的射频通道之间建立安全通道,传输加密的数据用于支付交易等。

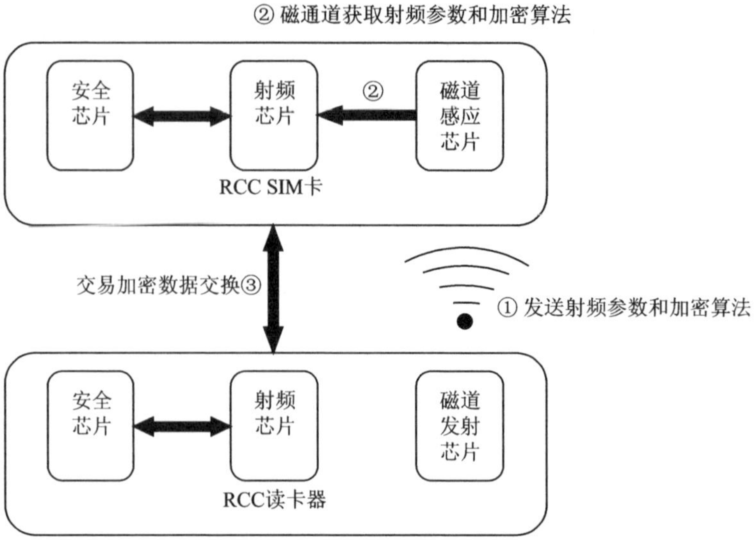

图 3-1 RCC 限域通信框图

目前,该技术主要以国民技术等国内企业为主要引擎推动 RCC 技术的发展。该技术于 2009 年成为中国移动企业标准,并于 2010 年率先在上海世博会应用——世博通,起草单位有中国信息通信研究院、国民技术股份有限公司、中国移动通信集团公司、中国联合通信有限公司、中国电信集团公司。根据国民技术的官方资料统计,到目前为止累计发行 RCC 卡超过 700 万张,深圳通 RCC 卡累计用户超过 170 万张。下面罗列几个 RCC 应用的重要时间节点,以及该技术在国内外主要的应用试点情况。

2009 年 7 月:RCC 移动支付产品通过中国移动应用测试;

2010 年 1 月:上海地铁开通基于 RCC 技术的手机刷卡功能;

2010 年 5 月:上海世博会开通基于 RCC 技术的手机票刷卡入园功能;

2011 年 1 月:阿塞拜疆开通基于 RCC 技术的手机支付业务;

2011 年 6 月:深圳全面开通基于 RCC 技术的手机深圳通业务;

2011 年 9 月:兰州开通基于 RCC 技术的手机支付业务;

2012 年 5 月:印尼开通基于 RCC 技术的手机支付业务;

2012 年 12 月:哈尔滨正式开通基于 RCC 技术的手机城市通卡业务;

2013年5月:青岛正式开通基于RCC技术的手机琴岛通业务;

2014年1月:包头正式开通基于RCC技术的手机鹿城通业务;

2014年1月:绵阳正式开通基于RCC技术的手机绵州通业务;

2017年5月:RCC限域通信技术正式颁布为国家标准GB/T 33736—2017。

RCC技术具有以下优势:第一,2.45 GHz射频频段以及低频准静态磁场对手机的相对适应性比较好;第二,距离控制可靠性高,采用低频准静态磁场控制距离,而非2.45 GHz频段本身。RCC在技术原理上保证了高安全性从而可应用于支付领域。低频准静态磁场辐射范围遵循物理规律(距离的增加、稳定的衰减、辐射范围极难人为延展),这一特性决定了RCC移动支付技术的安全性,其有着更强的抗攻击性能。RCC技术的高安全性主要体现在以下四方面:第一,基于物理规律的距离控制从原理上保障了高安全等级;第二,距离控制机制贯穿整个交易过程,保障了工程实施的高安全性;第三,高强度密码机制保障了链路层的高安全性;第四,采用独立的安全体系,不与支付终端硬件、操作系统挂钩,便于检测和进行系统安全评估。

3.1 RCC限域通信技术工作原理

RCC移动支付技术采用2.4 GHz射频通道和磁通道的双通道机制。磁通道与射频通道都含有协议层次模型、协议物理层、链路层、传输层和会话层所传输处理的数据单元,以及协议基本流程、协议防冲突机制和协议消息命令等。磁通道利用低频准静态磁场,以耦合方式完成可靠的距离控制;射频通道采用高带宽2.4 GHz频段,以电磁场发射接收方式完成高速数据交换;在物理层协议上,采用高强度密码技术将两个通道紧密融合,共同完成近场支付功能。RCC移动支付系统如图3-2所示。

支持RCC技术的卡耦合到磁信号后,解调解码出磁信息,然后根据磁信息动态生成2.45 GHz射频通道参数,使得卡与读卡器在此参数指定的动态改变的秘密信道上通信;磁信息承载于低频准静态磁场中,仅在安全范围内传播,能够保障2.45 GHz通道链路参数和数据的安全,是整个会话过程中通信双方进行安全通信的基础;交易过程中采用高强度密码算法可确保磁信道和2.45 GHz射频信道的紧密捆绑的高安全性。

RCC限域通信技术的两个前提条件为:第一,发起方在连接和交易过程中必须维持磁场存在,响应方在发送任何响应之前,必须确定其处于设定的磁场强度范围

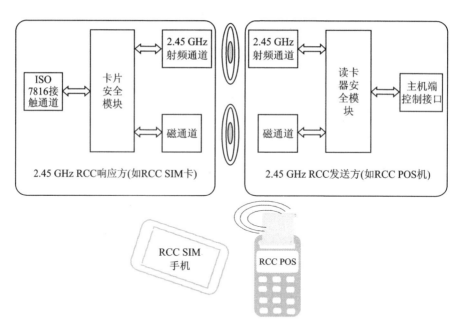

图 3-2 RCC 移动支付系统

内;第二,会话层消息的发送方在消息发送完成后应当立即转换为消息接收状态,并在会话命令规定的时间内进行接收。

RCC 协议会话的工作流程通常包括四个阶段:激活、接入、交易和结束,如图 3-3 所示为 RCC 通信的基本会话流程。

图 3-3 中实线箭头表示为射频通道,虚线箭头则为磁通道。

1. 激活阶段

1) 发起方操作

- 发起方在激活阶段通过磁通道发送 INQUIRY,然后通过 RF 通道接收 ATI。
- 发起方在发送 INQUIRY 之前,根据自己生成的 IDm 计算激活响应(ATI)频点并确认该频点是否可用,如果该频点当前已被占用,则需要重新生成 IDm 直到选择的 ATI 频点空闲为止。
- 发起方如果接收到错误的 ATI,或者 ATI 中的 Mac 验证失败,或者接收超时(>8 ms),则发送新的 INQUIRY。
- 发起方在接收到第一个正确的 ATI 之后进入接入阶段。

2) 响应方操作

- 响应方在激活阶段通过磁通道接收 INQUIRY,然后通过 RF 通道发送 ATI。

3 RCC 限域通信支付系统

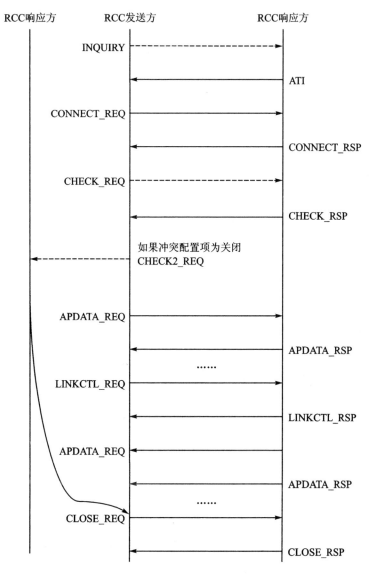

图 3-3 RCC 通信会话流程

- 响应方如果接收到错误的 INQUIRY 或者未接收到 INQUIRY，则继续接收 INQUIRY。
- 响应方在接收到第一个正确的 INQUIRY 之后被激活，然后在 8 ms 内通过 RF 通道做出 ATI 响应。
- 响应方在发送 ATI 之前，根据自己生成的 IDs 计算后续接入/交易频点并确认该频点是否可用，如果该频点当前已被占用，则需要重新生成 IDs 直到选

17

择的接入/交易频点空闲为止。
- 响应方在发送完 ATI 之后进入接入阶段。

2. 接入阶段

1) 发起方操作

- 发起方在接入阶段通过 RF 通道发出 CONNECT_REQ 连接请求,然后通过 RF 通道接收 CONNECT_RSP 响应。
- 发起方如果接收到错误的 CONNECT_RSP 或者接收超时(>8 ms),则返回激活阶段。
- 发起方在接收到第一个正确的 CONNECT_RSP 之后进入交易阶段。

2) 响应方操作

- 响应方在接入阶段通过 RF 通道接收 CONNECT_REQ,然后通过 RF 通道发送 CONNECT_RSP 进行响应。
- 响应方如果接收到错误的 CONNECT_REQ 或者接收超时(>8 ms),则返回到激活阶段。
- 响应方在发送完 CONNECT_RSP 之后进入交易阶段。

3. 交易阶段

交易阶段完成的操作可能会包含 5 种情况:数据交换、链路维持、冲突检测、连接确认和长时等待。

(1) 数据交换

1) 发起方操作

- 发起方在交易阶段通过 RF 通道发送 APDATA_REQ 数据交换请求,然后通过 RF 通道接收 APDATA_RSP 响应或者 LTW 长时等待消息。
- 发起方如果接收到错误的 APDATA_RSP/LTW 或者接收超时(>500 ms),则返回激活阶段。
- 发起方如果接收到正确的 APDATA_RSP/LTW,则继续维持交易阶段。

2) 响应方操作

- 响应方在交易阶段等待接收 APDATA_REQ,解析和执行封装在 APDATA_REQ 中的 APDU 命令,然后把对 APDU 命令的响应封装在 APDATA_RSP 中发送给发起方。

3 RCC 限域通信支付系统

- 响应方如果接收到错误的 APDATA_REQ 或者接收超时（>100 ms），则返回激活阶段。
- 响应方应该在 500 ms 内发送 APDATA_RSP，或者 LTW 长时等待消息给发起方。
- 响应方在发送完 APDATA_RSP 或 LTW 后，继续维持交易阶段。

（2）链路维持

1）发起方操作

- 发起方在交易阶段的空闲时间里应当每隔 44 ms 通过 RF 通道发送 LINKCTL_REQ，然后通过 RF 通道接收 LINKCTL_RSP 以维持连接。
- 发起方在发送 LINKCTL_REQ 后，如果在 8 ms 内接收到正确的 LINKCTL_RSP 响应且响应方状态正常，则继续维持与响应方的连接；如果发起方在连续 10 次发送 LINKCTL_REQ 后均无法正确收到 LINKCTL_RSP 响应，则认为响应方已断开连接，发起方返回激活阶段。

2）响应方操作

- 响应方在交易阶段，每次从 RF 通道接收到正确的 LINKCTL_REQ 命令请求时，均应在 8 ms 内发送 LINKCTL_RSP 做出响应。
- 响应方在交易阶段，如果接收到正确的 LINKCTL_REQ 命令请求，并确认磁通道已收到 CHECK1_REQ 或者 CHECK2_REQ 消息，且 CDC 或者 TRI 正确无误，则应当更新响应方连接状态为正常，并在 8 ms 内发送 LINKCTL_RSP 做出响应。
- 响应方在交易阶段，如果连续三次接收到正确的 LINKCTL_REQ 命令请求而没有收到任何 CHECK1_REQ 或者 CHECK2_REQ 消息，则应当更新响应方连接状态为异常，并在 8 ms 内发送 LINKCTL_RSP 做出响应。
- 响应方如果接收到错误的 LINKCTL_REQ，或者超过 100 ms 仍未接收到任何命令，则响应方返回激活阶段。

（3）冲突检测

当发起方检测到多响应方冲突发生时，发起方不能进行任意一个响应接入，且应结束与当前响应方的连接。多响应方冲突如图 3-4 所示。

1）发起方操作

- 发起方可以通过外部配置确定是否支持多响应方冲突检测。

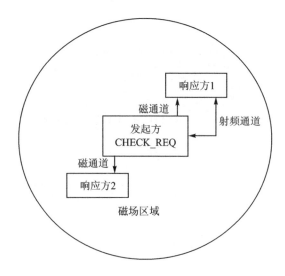

图 3-4　冲突检测示意图

- 支持冲突检测的发起方在建立连接后的整个交易阶段通过磁通道持续发送 CHECK1_REQ 消息。
- 发起方建立连接后,在射频通道空闲期间(没有维持连接和 APDU 收发任务时)以 t_2($t_2 \geqslant 4$ ms)为一个最小接收时间单元,持续在冲突响应信道上接收 CHECK_RSP 消息。
- 发起方如果接收到 CHECK_RSP,并且其中 TargetID 字段与当前连接的响应方 TargetID 不相符,则认为发生了多响应方冲突。
- 当发起方检测到多响应方冲突后,应当立即发送 CLOSE_REQ 指令,结束与当前响应方的连接。

2) 响应方操作

- 响应方如果在未连接状态下通过磁通道接收到 CHECK1_REQ 消息,则认为发生了多响应方冲突。
- 响应方两次收到 CHECK1_REQ 消息的最大时间间隔定义为一个冲突响应时间窗 T,本标准中 $T=22$ ms。
- 响应方检测到多响应方冲突后,在冲突响应时间窗中,以 t_1 为冲突响应时间间隔(t_1 最大为 4 ms)。t_1 被划分为 1 个或多个响应时隙(每个时隙 400 μs)。冲突响应方在 t_1 时间内随机选择一个时隙发送冲突响应消息。冲突检测处理时序如图 3-5 所示。

➢ 冲突响应消息发送频点和地址由 CHECK1_REQ 消息中的 CDC 字段决定。

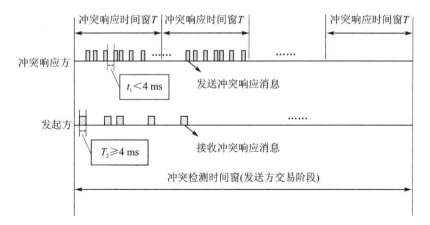

图 3-5 冲突检测处理时序图

（4）连接确认

1）发起方操作

➢ 如果冲突配置项为关闭，则发起方在整个交易阶段通过磁通道发送 CHECK2_REQ 连接确认请求。

2）响应方操作

➢ 如果响应方在交易阶段通过磁通道接收到 CHECK2_REQ 连接确认请求，则根据 TRI 是否正确，相应地更新响应方的当前连接状态。该状态信息将在响应方下一条 RF 响应消息中返回给发起方。

（5）长时等待

1）发起方操作

➢ 发起方在交易阶段等待响应方的 APDATA_RSP 命令响应期间，如果接收到来自响应方的正确的 LTW，则继续等待 500 ms，直到接收到 APDATA_RSP 或下一个 LTW，或者接收超时（>500 ms）。

➢ 发起方如果接收到错误的 LTW，则返回激活阶段。

2）响应方操作

➢ 如果响应方不能在一个 500 ms 时间内处理完成发起方的一个 APDATA_REQ 中包含的命令请求，则必须在 500 ms 内通过 RF 通道向发起方发送一个 LTW 长时等待消息，以通知发起方再等待 500 ms；如果响应方在下一个 500 ms 内仍不能处理完成发起方的 APDATA_REQ 命令请求，则继续在每个 500 ms 内发送一个 LTW，直到响应方处理完该交易后做出 APDATA_RSP 响应为止。

> 响应方发送完 LTW 之后，继续维持交易过程。

4. 结束阶段

1) 发起方操作

> 如果发生下列情况之一，发起方必须通过 RF 通道发出 CLOSE_REQ 关闭连接请求。

　　a) 交易正常结束；

　　b) 响应方状态不正常，即接收到的响应消息中的状态字段值不是 0x00；

　　c) 发生多响应方冲突，即接收到 CHECK_RSP。

> 发起方应当在 CLOSE_REQ 请求中对于是否要求响应方做出回应给出明确的指示。

> 如果发起方要求响应方回应，则继续等待，直到接收到 CLOSE_RSP 或接收超时（>500 ms）之后才返回激活阶段；否则发起方应在发送完 CLOSE_REQ 后立即返回激活阶段。

2) 响应方操作

> 响应方在接收到第一个正确的 CLOSE_REQ 后，应当立即结束交易，并根据 CLOSE_REQ 中的指示来决定是否发送 CLOSE_RSP 响应。

> 如果发起方要求回应，则响应方应当在 500 ms 内发送完 CLOSE_RSP，然后返回激活阶段；否则响应方应在接收到 CLOSE_REQ 后立即返回激活阶段。

3.2　RCC 限域通信协议综述

　　RCC 限域通信技术的协议从物理特性、数据链路、传输再到应用和业务接口，都有详细的定义和规范，此标准的协议也是根据短距离通信的特性，共划分为四层：物理层、数据链路层、传输层和会话层。如表 3-1 所列，其中应用层与具体的应用和业务相关，RCC 限域通信协议本身并不定义这个部分。

1. 物理层

　　物理层规定磁通道和射频通道的物理接口特性，包括磁通道的编码方式、调制方式、磁场强度要求,以及射频通道的频段和信道、调制方式、发射参数等物理特性。

表 3-1 RCC 限域通信协议族

层 数	模 型	接 口	示 例
5	应用层	应用和业务处理	APDU 数据包
4	会话层	消息	INQUIRY 激活命令
3	传输层	包	磁通道或者射频通道包
2	链路层	帧	磁通道或者射频通道帧
1	物理层	物理信号	定义磁通道或者射频通道的特性

2. 数据链路层

数据链路层规定磁通道和射频通道的帧格式、组帧和解帧,以及帧的发送和接收。定义链路层最小数据处理的单位为帧,链路处理行为均基于帧进行处理,对有效数据进行扩展,形成信道能够稳定传输的机制。

3. 传输层

传输层规定磁通道和射频通道的数据包格式、分包和组包,以及包的发送和接收。定义为传输层处理的最小数据单位为包,传输层处理行为均基于包进行处理,对有效帧进行扩展,从而形成批量数据的传输机制。

4. 会话层

会话层规定消息格式、消息功能定义、消息交互流程、应用与业务的接口。定义会话层处理的最小数据单位为消息,会话层处理行为均基于消息进行处理,对包进行扩展,并提供应用层的相关接口。

3.2.1 磁通道帧、包、消息数据单元

磁通道的数据单元分成两种类型:第一种为 Type1,若消息体为小于或等于 15 字节的数据包,则在传输层不做任何分包和组包处理,直接透传,如表 3-2 所列;第二种为 Type2,若消息体为大于 15 字节的数据包,则需要在传输层做分包和组包处理。

➢ 会话层的消息头和消息体,在传输层中只做透传处理。

➢ 会话层的消息头直接映射到链路层帧数据的控制域,消息体直接映射到帧数

据的数据域,另外加入 CRC 校验域,此种格式帧类型为 0000~1110b。

表 3-2 Type1-磁通道帧、包、消息单元

层 数	模 型	数据关系			
4	会话层	消息头	消息体(0~15字节)		
3	传输层				
2	链路层	控制域	数据域		CRC 校验域
		物理帧(对上面逻辑帧进行位填充＋同步码)			
1	物理层	〜〜〜〜〜〜〜〜〜〜			

➢ 物理层把链路层组装好的逻辑帧,通过 RCC 技术的磁通道编码和调制方式发送出去;或者反向根据差分曼彻斯特编码原理把数据解析回来后,送到链路层进行 CRC 校验处理,无误后解析出控制域和数据域,通过传输层透传给会话层,解析出消息头和真正的消息体。

第二种数据单元,当消息体大于 15 字节后,就需要在传输层进行分包和组包处理了。表 3-3 所列为当会话层准备发送 448 字节时,它们在各层之间传递的数据帧、包、消息的关系。

表 3-3 Type2-磁通道帧、包、消息单元

层 数	模 型	数据关系				
4	会话层	消息体(448字节)				
3	传输层	数据包 0 (14字节＋0)	数据包 1 (14字节＋1)	数据包 2 (14字节＋2)	(14字节＋n＜32)	数据包 32 (14字节＋32)
		数据包 0(14 字节＋0)				
2	链路层	控制域	数据域			CRC 校验域
		物理帧(对上面的逻辑帧进行位填充＋同步码)				
1	物理层	〜〜〜〜〜〜〜〜〜〜				

➢ 会话层准备发送一次消息体为 448 字节的通信。
➢ 当传输层发现 15 字节的数据包无法一次性传递完时,传输层会对会话层传递的消息体进行分包处理,对消息体数据除以 14 字节,算出需要分包多少次进行传输,如果有余数则需要在最后发送一包所剩余数的包。这种分包处理过的包会在 14 字节后,把分包的序列号拼成一个字节加在后面,此种格式帧

类型为1111b。
- 物理层把链路层组装好的逻辑帧,通过RCC技术的磁通道编码和调制方式发送出去;或者反向根据差分曼彻斯特编码原理把数据解析回来后,送到链路层进行CRC校验处理,无误后解析出控制域和数据域,通过传输层进行组包处理后再传给会话层,解析出真正的消息体。

磁通道数据帧格式

磁通道数据链路层帧分为逻辑帧和物理帧,逻辑帧定义了磁通道数据的逻辑结构,采用长编码格式,其帧结构包括:控制域(包括帧类型和帧数据长度)、数据域和CRC校验域;通过对逻辑帧数据进行位填充并增加同步码,就形成了真正在磁通道上传输的物理帧。如图3-6所示为磁通道数据帧格式。

逻辑帧			
控制域		帧数据域	CRC校验
帧类型	帧数据长度	帧数据	
ToF	DataLen	MC_Data	CRC-8
4比特	4比特	0~15字节	1字节
物理帧			
同步码	逻辑帧+位填充码		
1111 1111 0b	逻辑帧从起始向后检索,当出现1111 111b时,填充1位"0"形成1111 1110b		

图3-6 磁通道数据帧格式

(1) ToF(帧类型)

ToF为帧类型,用于标识不同类型的磁通道帧,包括直接用于传输磁通道的基本帧,或者用于传输磁通道的扩展帧,长度为4比特。

- 0000b~1110b:Type1 磁基本帧;
- 1111b:Type2 磁扩展帧。

(2) DataLen(帧长度)

DataLen表示帧数据域中的字节长度,为4比特,取值范围为0000b~1111b。

(3) MC_Data(磁道里的帧数据域)

MC_Data表示需要传输的磁道里的帧数据域,此域中的数据长度由DataLen字段值定义。

(4) CRC-8(帧校验域)

计算出帧结构中的控制域和数据域总的CRC校验码,长度为1字节,取值范围为 0x00～0xFF。CRC-8 的计算方法为：① 8 位 CRC 校验初始值为 0x00；② 8 位 CRC 校验的多项为 X^8+X^2+X+1。

3.2.2 射频通道帧、包、消息数据单元

射频通道的数据单元与磁通道的 Type2 方式比较相似，只是在传输层组包和分包时采用的字节大小为 32 字节。表 3-4 所列为当会话层准备发送 992 字节时，它们在各层之间传递的数据帧、包、消息的关系。

表 3-4 射频通道帧、包、消息单元

层数	模型	数据关系				
4	会话层	消息体(992 字节)				
3	传输层	数据包 0 (31 字节+0)	数据包 1 (31 字节+1)	数据包 2 (31 字节+2)	31 字节+n(<32)	数据包 32 (31 字节+32)
2	链路层	数据包 0 (31 字节+0)				
2	链路层	控制域	数据域		CRC 校验域	
1	物理层	～～～～～～				

- 会话层准备发送一次消息体为 992 字节的通信。
- 当传输层发现 32 字节的数据包无法一次性传递完时，传输层会对会话层传递的消息体进行分包处理，对消息体数据除以 31 字节，算出需要分包多少次进行传输，如果有余数则需要在最后发送一包所剩余数的包。这种分包处理过的包会在 31 字节后，把分包的序列号拼成一个字节加在后面。
- 物理层把链路层组装好的逻辑帧，通过 RCC 技术的射频通道编码和调制方式发送出去；或者反向根据射频通道编码原理把数据解析回来后，送到链路层进行 CRC 校验处理，无误后解析出控制域和数据域，通过传输层进行组包处理后再传给会话层，解析出真正的消息体。

射频通道数据帧格式

射频通道数据链路层帧格式采用变长编码格式，其帧结构包括：前导码、地址、控制域(包括数据长度、帧标识、应答标识)、数据域、CRC 校验，如表 3-5 所列。

(1) RF_Preamble 前导码

用于射频通道数据帧同步，长度为 1 字节。

- 01010101b：如果地址 Address 域最高位为 0，则 RCF 前导码为 01010101b；
- 10101010b：如果地址 Address 域最高位为 1，则 RCF 前导码为 10101010b。

表 3-5 射频通道数据帧格式

前导码	地址	控制域			数据域	CRC 校验域
		数据长度	帧标识	应答标识		
RF_Preamble	Address	RF_DataLen	FrameID	AckFlag	RF_Data	CRC-16
1 字节	40 比特	6 比特	2 比特	1 比特	0~32 字节	16 比特

（2）Address 地址

用于射频通道数据帧接收的识别地址，长度为 5 字节(40 比特)。

取值范围为 0x0000000000~0xFFFFFFFFFF。

（3）RF_DataLen 数据长度

标识射频通道数据帧中数据域 RF_Data 的字节长度，长度为 6 比特。

取值范围为 0x000000~0x100000。

当 RF_DataLen 数据长度为 0 时，仅限于 RF 通道的 ACK 应答。

（4）FrameID 帧标识

用于区分不同的射频通道数据帧，相邻的不同的帧标识应当不同，长度为 2 比特。

取值范围为 00b~11b。

（5）AckFlag 应答标识

用于射频通道数据帧接收方判断是否需要发送 ACK 应答，长度为 1 比特。

- 0b：接收方不发送 ACK 应答(当 RF_DataLen 为 0，且无 RF_Data 数据时，也叫数据帧)。
- 1b：接收方自动发送 ACK 应答(也叫应答帧)。

（6）RF_Data 帧数据

表示射频通道数据帧中传输的数据域，长度由 RF_DataLen 字段定义。

（7）CRC-16 校验

计算射频通道数据帧中，除 RF_Preamble 前导码和 CRC-16 本身之外所有的域，CRC-16 的计算方法为：① 16 位 CRC 校验初始值为 0xFFFF；② 16 位 CRC 校验的多项式为 $X^{16}+X^{12}+X^5+1$。

3.3 RCC 限域通信射频通道帧收发

按照前面介绍的射频通道帧格式,将数据组织在一起,再经过物理层处理后形成射频信号,并通过 2.45 GHz 频率发送出去。发送的顺序依次为:① 发送 RF_Preamble 前导码,确定成功;② 发送 Address 地址域,确定成功;③ 发送控制域,确定成功;④ 发送数据域,确定成功;⑤ 发送 CRC-16 校验域,确定成功;⑥ 结束,上面的全部过程也简称为组帧。

在物理层将接收的射频信号解调成数字比特信号后,按照帧结构进行解析,接收的顺序依次为:① 接收 RF_Preamble 前导码,确定成功;② 接收 Address 地址域,确定成功;③ 接收控制域,确定成功;④ 接收数据域,确定成功;⑤ 接收和计算 CRC-16 校验域,确定成功;⑥ 结束,解出有效的帧数据的过程也简称为解帧。

射频帧的收发时序如图 3-7 所示。

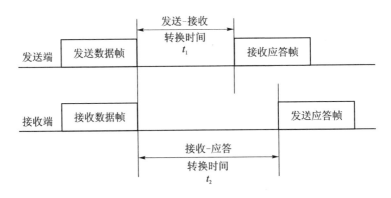

图 3-7 射频帧的收发时序

(1) t_1 发送-接收转换时间

发送方在完成数据帧发送后,必须在 t_1 时间内切换到应答帧接收状态($t_1 \leqslant 130\ \mu s$)。

(2) t_2 接收-应答转换时间

接收方在收到一个数据帧后,必须在 t_2 时间内发送一个应答帧进行响应($130\ \mu s < t_2 < 150\ \mu s$)。

(3) 帧传输

帧的传输过程包括一个数据帧发送和一个应答帧接收。发送方发送一个数据帧,并在 t_1 时间内切换到应答帧接收状态,如果成功接收到应答帧,则判断一次帧传输成功;如果未接收到应答帧,则判断帧传输失败;接收方在收到相同的数据帧后,应当丢弃并继续接收。

3.4 RCC 限域通信技术的物理层特性定义

3.4.1 磁通道 MC 的物理特性

磁通道用于 RCC 发送方和响应方之间进行通信距离限制和数据传输。发送方发射出经过编码和调制过的磁场信号,响应方检测磁场信号强度,从而实现通信距离的限制,并对接收到的磁场信号进行解调和解码,从而实现磁通道数据传输功能。下面将对磁通道通信的物理电气特性和编码调制方式进行介绍。

1. 数据编码符号率

磁通道数据编码符号率为 4 kS/s,符号率容许偏差范围为 ±5%。

2. 调制方式

磁通道信号采用磁场强度变化率调制,如图 3-8 所示。

- 符号"1"的场强变化率与符号"0"的场强变化率应为相反的关系,且变化率大小相等并保持恒定。
- 磁通道数据编码符号率为 4 kS/s,每个符号周期为 250 μs,H_p 为发起方磁场信号强度峰值。
- 磁通道信号符号的极性只与磁场强度变化率的方向有关,而与磁场强度变化率的大小无关。
- 磁场强度变化率的大小决定了磁场强度的峰值,磁场强度峰值的规范要求用于距离控制。
- 磁通道数据编码采用差分曼彻斯特编码(DME),具体细节如图 3-8 所示。
- 发起方根据上述编码和调制原理,将磁通道数据转换为磁场强度变化率调制信号。响应方通过磁感应线圈感应得到解调耦合电压信号,电压信号与磁场信号的关系如下:

$$V = \frac{N\mathrm{d}\varphi}{\mathrm{d}t} = \frac{N\mathrm{d}(B \cdot S)}{\mathrm{d}t} = \frac{N \cdot S\mathrm{d}B}{\mathrm{d}t} = \frac{N \cdot S\mathrm{d}(\mu_0 H)}{\mathrm{d}t} = \mu_0 \cdot N \cdot S \frac{\mathrm{d}H}{\mathrm{d}t}$$

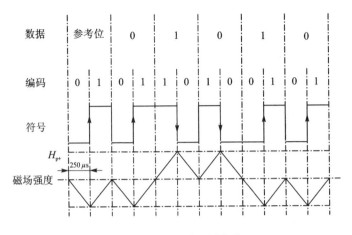

图 3-8 磁通道调制方式

式中:V 为响应方的耦合电压;N 为响应方耦合线圈的匝数;S 为响应方耦合线圈的面积;$\dfrac{dH}{dt}$ 为磁场强度的变化率;μ_0 为空气中的磁导率。

> 响应方从电压信号中得到磁通道信号符号序列,再根据差分曼彻斯特编码机制解码得到磁通道数据,根据差分曼彻斯特编码机制中参考位和数据位的规则,不管响应方采用的是正向耦合方式还是反向耦合方式,最终都是可以获取到正确的解码数据。

3. 编码方式

磁通道数据编码采用差分曼彻斯特编码(DME),如图 3-9 所示。

每个数据位由 2 个符号组成的序列表示,每个数据位的符号序列必须为"10"或"01"。数据位"1"的符号序列与前一个数据位的符号序列相反,数据位"0"的符号序列与前一个数据位的符号序列相同

4. 发起方磁场信号强度

以发起方设备工作位置中心为基准,工作方向距设备表面垂直距离 0 cm 处磁场信号强度峰值应不小于 160 A/m,且不大于 300 A/m;距设备表面垂直距离 10 cm 处磁场信号强度峰值应小于 4.2 A/m。

5. 响应方磁场信号强度

> 响应方设备置于发起方设备的磁场信号工作区域,在磁场信号强度峰值 H_p 不小于 6.7 A/m 时,响应方应能够与发起方建立并保持连接。

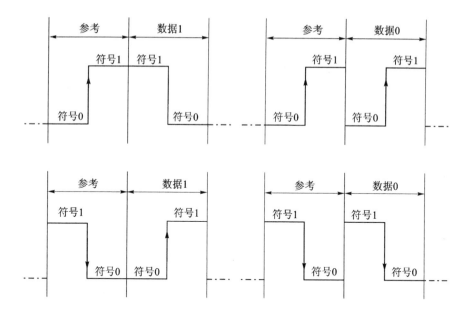

图 3-9 磁通道编码方式

➢ 响应方设备置于发起方设备的磁场信号工作区域,在磁场信号强度峰值 H_p 不大于 4.2 A/m 时,禁止响应方与发起方建立或保持连接。

6. 磁场信号符号周期抖动

磁场信号各个符号周期对理想值(250 μs)的偏离范围为磁场信号符号周期抖动,磁场信号符号周期抖动应不大于 60 μs。

3.4.2 射频通道 RC 的物理特性

射频通道采用 2.45 GHz 工业、科学和医疗(ISM)频段,以电磁场发射接收方式完成高速数据交换。协议会话层采用密码技术对射频通道交换的应用协议数据单元(APDU)的数据进行加密传输。

1. 射频频段和信道分配

射频通道的工作频率范围为 2 400～2 483.5 MHz,射频信道间隔为 1 MHz,各信道标称中心频率如表 3-6 所列。

表 3-6 射频频段范围和信道中心频率

射频频段范围 /MHz	射频信道标称中心频率 f_c/MHz
2 400～2 483.5	$f_c = 2\ 400 + k$, $k = 1, 2, \cdots, 83$

信道的划分如表 3-7 所列。

表 3-7 射频信道划分

2 400 MHz	2 401 MHz	…	2 464 MHz	2 465 MHz	…	2 468 MHz	2 469 MHz	…	2 483 MHz
保留	近距离通信			冲突检测			保留		

2. 射频发射功率

各射频信道传导发射功率最大不得超过 +3 dBm,有效全向辐射功率(EIRP)不得超过 10 mW。

3. 射频频率容限

发射载波的初始中心频率 f_t 应在本信道标称中心频率 f_c 的 ±75 kHz 频率范围内,即 $f_c-75\ \text{kHz} \leqslant f_t \leqslant f_c+75\ \text{kHz}$。注意:±75 kHz 不包括数据发送过程中的频率漂移。在一个数据帧传输时间内,发射载波中心频率累积漂移量应在 ±20 kHz 之内。

4. 调制参数

射频通道信号调制方式为高斯移频键控(GFSK),带宽与码元宽度的乘积参数 BT 为 0.5,符号率为 1 MS/s,调制指数应在 0.27～0.55 之间,即频率偏移幅度应在 135～275 kHz 之间。数据位"1"以正频偏表示,数据位"0"以负频偏表示。在任何情况下,最小频率偏移不应小于 115 kHz。如图 3-10 所示,在 1010 数据序列传输中对应的最小频率偏移幅度 $f_{\min}$ 应不小于 ±80% 的 00001111 数据序列传输中的频率偏移幅度 f_d。

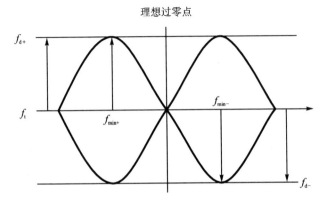

图 3-10 射频通道信号眼图

5. 杂散辐射

在 RCC 技术的射频通道上，发送方的杂散辐射是指用标准信号调制时在除载频和由于正常调制和切换瞬态引起的边带以及邻道以外离散频率上的辐射，表 3-8 为带内杂散功率应符合的最大值要求，其中第 M 信道为射频信号发射信道，第 N 信道为相邻信道。

表 3-8 发射频谱带内杂散

相邻信道间隔	杂散功率最大限值
2 MHz($\|M-N\|=2$)	−20 dBm
⩾3 MHz($\|M-N\|⩾3$)	−30 dBm

带外杂散应符合表 3-9 所列的频率范围、测试带宽、绝对值等要求。

表 3-9 发射频谱带外杂散

频率范围	测试带宽(RBW)	限值(绝对值)	检波方式
30 MHz~1 GHz	100 kHz(3 dB)	−36 dBm	有效值
1~12.75 GHz	1 MHz(3 dB)	−30 dBm	有效值

3.5　RCC 限域通信技术的会话层命令定义

本节把协议会话工作流程中包括的激活、接入、交易和结束四个阶段中所用到的全部会话命令集合为表 3-10。会话命令是通过发起方和响应方之间传递消息的方式来实现的，下面会对各个命令数据体进行详解。注意本节中使用的"xx"代表数据无确定值或者属于随机数。

表 3-10 会话层命令定义

类型	消息码	名称	功能	信道
激活	0	INQUIRY	激活请求	磁通道
	16	ATI	激活响应	
连接	17	CONNECT_REQ	连接请求	射频通道
	18	CONNECT_RSP	连接响应	
数据交换	19	APDATA_REQ	数据交换请求	
	20	APDATA_RSP	数据交换响应	

续表 3-10

类　型	消息码	名　称	功　能	信　道
连接维持	22	LINKCTL_REQ	维持连接请求	射频通道
	23	LINKCTL_RSP	维持连接响应	
冲突检测	2	CHECK1_REQ	冲突检测请求和响应，可适用于冲突检测和连接确认	磁通道
	24	CHECK_RSP		射频通道
连接确认	3	CHECK2_REQ	连接确认请求，冲突检测关闭时用于连接确认	磁通道
长时等待	25	LTW	响应方要求等待	
关闭	26	CLOSE_REQ	关闭连接请求	射频通道
	27	CLOSE_RSP	关闭连接响应	

3.5.1 命令交互的频点和地址

发起方在发送 INQUIRY 命令之前，会根据自己生成的 IDm 计算激活响应 ATI 的射频频点并确认该频点是否可用，如果该频点当前被占用，则需要重新生成 IDm 直到选择到激活响应 ATI 的频点为空闲可用为止。

当发送方和响应方成功协商出频点和地址数据后，之后的连接、数据交换、维持连接、冲突检测、长时等待和关闭等会话命令都会基于这个频点和地址的算法进行数据交换，表 3-11 为会话命令集使用的频点和地址汇总。

表 3-11 会话命令集频点和地址汇总

频点和地址	命　令
freq1,addr1	ATI
freq1,addr2	CONNECT_REQ,CONNECT_RSP,APDATA_REQ, APDATA_RSP,LINKCTL_REQ,LINKCTL_RSP, LTW,CLOSE_REQ,CLOSE_RSP
freq2,addr1	CHECK_RSP
freq2,addr2	N/A

1. 频点的计算方法

以 N 为模对输入 X 进行求余运算，即 (X) Mod N，其中 N 是射频频率表支持的最大频点数目，X 为不小于 2 字节的数据串。表 3-12 为频点对应关系。

2. freq1 算法描述

用途:计算工作频点。

表 3 – 12　频点对应关系

频点序号	Ch_1	Ch_2	Ch_3	Ch_4	…	Ch_{N-1}	Ch_N
余数	0	1	2	3	…	N−2	N−1
工作频率/MHz	2 401	2 402	2 403	2 404	…	2 400+N−1	2 400+N
冲突响应频率/MHz	2 465	2 466	2 467	2 468	—	—	—

计算方法:

① X 为不小于 2 字节的数据串,取 X 前 2 字节,记为 X2。

② freq1(X)=(X2) Mod N,N=64。

③ 用 freq1(X)作为索引,查表 3 – 12 得到工作频点。

计算示例:

假设数据串 X 为 0x30 ‖ 0x39 ‖ 0x23 ‖ 0xa5,其中" ‖ "表示"拼接"。

X2=0x30 ‖ 0x39=12345。

freq1(X)=(X2) Mod N=(12345) Mod 64=57,根据余数与频点的对应关系,余数 57 对应频点 Ch_{58},查表 3 – 12,Ch_{58} 频率值为 2 458 MHz。

3. freq2 算法描述

用途:计算冲突响应(CHECK_RSP)频点。

计算方法:

① X 为不小于 2 字节的数据串,取 X 前 2 字节,记为 X2。

② freq2(X)=(X2) Mod N,N=4。

③ 用 freq2(X)作为索引,查表 3 – 12 得到冲突响应频点。

计算示例:

假设数据串 X 为 0x30 ‖ 0x39 ‖ 0x23 ‖ 0xa5,其中" ‖ "表示"拼接"。

X2=0x30 ‖ 0x39=12345。

freq2=(X) Mod N=(12345) Mod 4=1,根据余数与频点的对应关系,余数 1 对应频点 Ch_2,查表 3 – 12,Ch_2 频率值为 2 466 MHz。

4. addr1 算法描述

用途:计算激活响应(ATI)和冲突响应(CHECK_RSP)射频通信地址。

计算方法:

addr1(X)=X[0]‖X[1]‖X[0]‖X[1]‖0x00,其中"‖"表示"拼接"。

X 为 2 字节数据串。

说明:按照下列顺序将各字节组合成一个 5 字节的 RF 地址,X 第 1 字节→X 第 2 字节→X 第 1 字节的按位取反→X 第 2 字节的按位取反→最后一个字节补 0。

5．addr2 算法描述

用途:计算除了 ATI 地址和 CHECK_RSP 地址之外的其他所有 RF 通信地址。

计算方法:

addr2(X)=X[0]‖X[1]‖X[2]‖X[3]‖X[4],其中"‖"表示"拼接"。

X 为 5 字节数据串。

说明:按照下列顺序将 X 各字节组合成一个 5 字节的 RF 地址,X 第 1 字节→X 第 2 字节→X 第 3 字节→X 第 4 字节→X 第 5 字节。

6．AID 计算方法

对于长度为 $n(2 \leqslant n \leqslant 14)$ 字节的 IDm,经过如下方式计算得到 AID:

① 若 $n \leqslant 8$,则在 IDm 后补 0x00($n=8$ 时不用补 0x00),补齐 8 字节,作为 Ka,K=Ka‖Ka(其中 Ka 表示对 Ka 的按位取反);若 $8 < n \leqslant 14$,则直接在 IDm 后补 0x00,补齐 16 字节,作为 K。

② 计算 3DES[K,IDm](若 $n<8$,则在 IDm 后补 0x00,补齐 8 字节作为被加密明文;若 IDm 长度 $n \geqslant 8$,则取 IDm 前 8 字节作为被加密明文),得到 8 字节密文数据。

③ 取 8 字节密文数据前 2 字节,作为 AID。

3.5.2　命令消息格式

在 2.45 GHz RCC 限域通信的发送方与响应方之间发送的磁通道数据或者射频通道数据,都需要遵循两种消息的格式,分别是短消息格式(见表 3-13)和长消息格式(见表 3-14)。短消息格式主要用于传输磁通道基本消息,而长消息格式主要用于传输射频通道消息和磁通道扩展消息(在磁通道上传输不超过 15 字节长度的消息,称为磁通道基本消息,大于 15 字节长度的消息称为磁通道扩展消息)。

表 3-13 中的消息码,用于标识某一类的短格式的消息;消息长度,用于表示消息体的字节长度;消息体中的有效数据域,用于承载消息所要传递的有效数据体。

表 3-13　短消息格式

消息头		消息体
消息码	消息长度	有效数据域
MsgCode	MsgLen	MsgData
4 比特(0000b～1111b)	4 比特(0000b～1111b)	0～15 字节(MsgLen 字段定义)
最高有效位		最低有效位

表 3-14　长消息格式

消息头					消息体	消息校验
保留	消息类型	状态	消息码	消息长度	数据域	校验体
Rfu	Format	Status	MsgCode	MsgLen	MsgData	CheckSum
4 比特	4 比特	8 比特	8 比特	16 比特	xx 字节	2 字节
最高有效位						最低有效位

表 3-14 中各个字段的说明如下：

- Rfu 保留位，长度为 4 比特，默认为 0000b；
- Format 消息类型，长度为 4 比特，表示消息的编码格式类型，其中 000b～0111b 表示私有格式，1000b 表示本编码格式，1001b～1111b 为保留位；
- Status 状态，长度为 8 比特，代表发送方或者响应方的当前状态，其中 0x00 表示正常，0x01 代表连接异常，0x02 表示响应方在射频发送时移出了允许的交易距离范围，0x82 指响应方在射频接收时移出了允许的交易距离范围，0xD0～0xFF 为自定义状态，其他为保留；
- MsgCode 消息码，长度为 8 比特，用于标识某一个具体的消息码；
- MsgLen 消息长度，长度为 16 比特，表示消息体的字节长度，不包含消息校验的数据体字节；
- MsgData 消息数据体，长度由 MsgLen 字段值定义，用于承载消息所要传递的有效数据体；
- CheckSum 消息校验，长度为 2 字节，用于对消息头和消息体数据进行 CheckSum 校验。

3.5.3　INQUIRY 激活命令

INQUIRY 激活命令，为发起方通过磁通道发给响应方的查询和激活命令，消息格式为短消息编码,命令格式如表3-15所列。

表3-15 INQUIRY激活命令

格式	标签	长度	值域	含义
消息头	MsgCode	4比特	0	消息码
	MsgLen	4比特	15	消息体的字节长度
消息体	Rfu	4比特	xx	保留域,默认为0
	InitiatorVersion	4比特	xx	发起方协议版本号,本书参考的协议版本号为0x03
	IDm	14字节	xx	发起方生成的随机数,用于计算ATI接入参数和后续的会话密钥

3.5.4 ATI激活响应命令

ATI激活响应命令,响应方通过射频通道的频点freq1 AID和地址addr1 AID进行数据传输。消息格式为长消息编码,命令格式如表3-16所列。

表3-16 ATI激活响应命令

格式	标签	长度	值域	含义
消息头	Rfu	4比特	xx	保留域,默认为0
	FormatType	4比特	8	消息格式类型
	Status	8比特	xx	响应方的当前状态
	MsgCode	8比特	16	消息码
	MsgLen	16比特	24	消息体字节长度
消息体	IDs	5字节	xx	响应方产生的随机数,用于双方计算通信的参数
	TargetID	8字节	xx	响应方唯一标识
	AccessVersion	1字节	xx	响应方将INQUIRY中的InitiatorVersion值与自身协议版本相比较,并且将双方同时支持的最高协议版本通过AccessVersion域返回给发起方
	MacData	4字节	xx	响应方使用K0作为密钥对(IDs ‖ TargetID ‖ AccessVersion)进行MAC计算得到MacData;16字节K0用SKG0中同样的扩展方式,由INQUIRY中14字节的IDm扩展生成
	Reserved	6字节	xx	保留域,默认为0
校验体	CheckSum	2字节	xx	校验值

MAC 算法

MAC 算法产生的步骤如下:

第一步:将一个长 8 字节的初始值(Initial Vector)设为十六进制的 0x 00 00 00 00 00 00 00 00。

第二步:将所有的输入数据按指定顺序连接成一个数据块。

第三步:将连接成的数据块分割为长 8 字节的数据块组,标识为 D_1、D_2、D_3、D_4 等。分割到最后,余下的字节组成一个长度小于或等于 8 字节的最后一个数据块。

第四步:如果最后一个数据块长度为 8 字节,则在此数据块后附加一个长 8 字节的数据块,附加的数据块为十六进制的 0x 80 00 00 00 00 00 00 00。如果最后一个数据块长度小于 8 字节,则在该数据块的最后填补一个值为十六进制的 0x80 字节。如果填补之后的数据块长度等于 8 字节,则跳至第五步。如果填补之后的数据块长度仍小于 8 字节,则在数据块后填补十六进制的 0x00 字节至数据块长度为 8 字节。

第五步:MAC 的产生是通过上述方法产生的数据块组,由 K0 进行加密运算实现的,加密算法 DEA 使用 DES。

第六步:最终值的左 4 字节为 MAC。MAC 算法如图 3-11 所示。

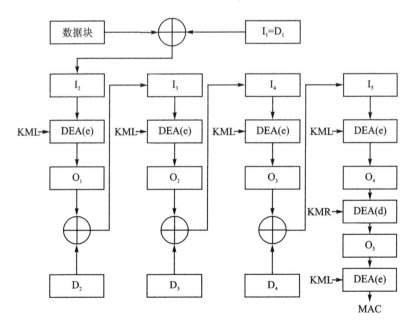

图 3-11　MAC 算法

3.5.5 CONNECT_REQ 连接请求命令

CONNECT_REQ 连接请求命令,为发起方通过射频通道的频点 freq1 IDs 和 addr2 IDs 进行数据传输,发起方在发送 CONNECT_REQ 消息之前,必须首先验证 ATI 中的 Mac 是否正确。发起方必须在 CONNECT_REQ 消息中指明本次连接中所希望采用的链路安全机制,包括使用的根密钥、会话密钥生成方式,以及加密算法等。消息格式为长消息编码,命令格式如表 3-17 所列。

表 3-17 CONNECT_REQ 连接请求命令

格式	标签	长度	值域	含义
消息头	Rfu	4 比特	xx	保留域,默认为 0
	FormatType	4 比特	8	消息格式类型
	Status	8 比特	xx	发起方的当前状态
	MsgCode	8 比特	17	消息码
	MsgLen	16 比特	24	消息体字节长度
消息体	InitiatorType	1 字节	xx	发起方类型: "A":近距离(10 cm); 其他:保留
	InitiatorID	8 字节	xx	发起方唯一标识
	RootKeyIndex	1 字节	xx	链路 APDU 数据加密的根密钥索引号,最多可支持 256 组根密钥。 0:表示选择 IDm 作为动态根密钥 K0 的有效密钥位,IDm 经位扩展后得到 16 字节 K0; 1:表示选择第 1 组预置根密钥; 2:表示选择第 2 组预置根密钥; $n<255$:表示选择第 n 组预置根密钥; 255:表示选择第 255 组预置根密钥
	SessionKey	1 字节	xx	发起方支持的会话密钥生成方式(SeesionKeyGenerate SKG): 每一比特代表一种会话密钥的生成方式,0 表示不支持该方式,1 表示支持该方式,最少应支持 1 种会话密钥生成方式,最多可支持 8 种会话密钥生成方式。 默认的会话密钥生成方式为 SKG0

续表 3-17

格式	标签	长度	值域	含义					
消息体	SessionKey	1字节	xx	位	值	生成方式	含义		
				b7	0	SKG7	保留位,默认为0		
				b6	0	SKG6	保留位,默认为0		
				b5	0	SKG5	保留位,默认为0		
				b4	0	SKG4	保留位,默认为0		
				b3	0	SKG3	保留位,默认为0		
				b2	0	SKG2	保留位,默认为0		
				b1	0	SKG1	保留位,默认为0		
				b0	1	SKG0	参考SKG0会话密钥产生方法		
	EncAlg	2字节	xx	发起方支持的链路加密算法模式:每一比特代表一种链路加密算法模式,0表示不支持该算法模式,1表示支持该算法模式,最少应支持1种加密算法模式,最多可支持16种加密算法模式。默认的链路加密算法为ALG0					
				位	值	算法模式	含义		
				b7~b15	0	N/A	保留位默认为0		
				b5	0/1	SM4-CBC	参见第2章 术语和缩略语		
				b4	0/1	SM4-ECB			
				b3	0/1	AES-CBC			
				b2	0/1	AES-ECB			
				b1	0/1	3DES-CBC			
				b0	0/1	3DES-ECB			
	MDInfo	5字节	xx	发起方设备信息,由厂商自定义					
	Reserved	6字节	xx	保留域,默认为0					
校验体	CheckSum	2字节	xx	校验值					

1. SKG0 会话密钥产生方法

第一步:计算主密钥 K,如果使用预置根密钥(即 CONNECT_RSP 中确认的 RootKeyIndex 值不为0),则 16 字节主密钥 K=K0⊕Ki,(i≠0);如果不使用预置根密钥(即 CONNECT_RSP 中确认的 RootKeyIndex 值为0),则 16 字节主密钥 K=K0,其中,16 字节 K0 由 14 字节 IDm(随机数)通过如图 3-12 所示方式扩展生成。

第二步:以响应方产生的 8 字节随机数 SDRand 为分散参数 X,按照下面定义的"子密钥分散算法"对第一步中计算得到的主密钥 K 进行密钥分散,得到 16 字节子

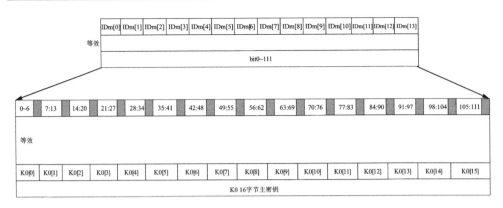

图 3-12　SKG0 会话密钥产生方法

密钥 Ks。

第三步：使用 Ks 作为会话密钥。

子密钥分散算法如图 3-13 所示，分散参数 X 若不足 8 字节，则先右补 0x80；若还不足 8 字节，则补 0x00 至 8 字节；若 X 超过 8 字节，则取最右边的 8 字节。

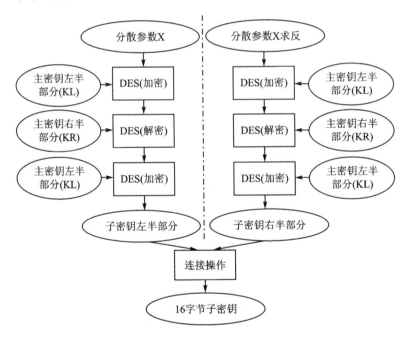

图 3-13　子密钥分散算法

2. 3DES 加/解密算法

3DES 加/解密算法定义如下：密钥长度为 16 字节（K=(KL‖KR)），数据分组长度为 8 字节。

对于每个数据分组的加密算法如下：

Y＝3DES[K,X]＝DES(KL)[DES－1(KR)[DES(KL)[X]]]

解密算法如下：

Y＝3DES－1[K,X]＝DES－1(KL)[DES(KR)[DES－1(KL)[X]]]

每个分组的加/解密过程如图 3－14 所示，对于多个分组的加/解密过程，只需要将所有加/解密后的数据块依照原顺序连接在一起即可。

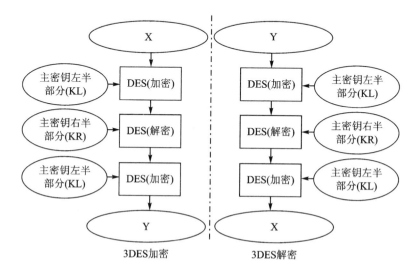

图 3－14　加/解密过程

3. ALG0 数据报文加密方式

对应用层数据报文数据体采用 ECB 模式的对称密码算法进行加/解密。

（1）按照如下步骤对数据报文进行加密

第一步：用 PLen(2 字节)表示明文数据的字节长度，在明文数据前加上 PLen，产生新的数据块。

第二步：将该数据块分成以分组长度 8 字节为单位的数据块，表示为块 1，块 2，…，块 n。

第三步：如果最后(或唯一)的数据块长度是分组长度，则转到第四步；如果没有达到分组长度，则在其后加入十六进制数 80；如果达到分组长度，则转到第四步；否则在其后加入十六进制数 00 直到长度达到分组长度。

第四步：按照图 3－15 所示的算法使用加密密钥对每一个数据块进行加密。

第五步：计算结束后，所有加密后的数据块依照原顺序连接在一起。

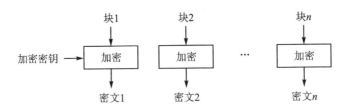

图3-15 数据报文加密步骤

（2）按照如下步骤对数据报文进行解密。

第一步：将该数据块分成以分组长度8字节为单位的数据块，表示为块1，块2，…，块n。

第二步：按照图3-16所示的算法使用解密密钥对每一个数据块进行解密。

第三步：计算结束后，所有解密后的数据块依照原顺序连接在一起。

第四步：前2字节为PLen，从第3字节起，取前PLen字节数据作为明文输出。

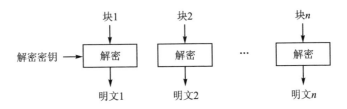

图3-16 数据报文解密步骤

3.5.6 CONNECT_RSP 连接响应命令

CONNECT_RSP 连接响应命令，响应方通过射频通道的频点 freq1(IDs) 和地址 addr2(IDs) 进行数据传输，响应方必须在 CONNECT_RSP 消息中确认本次连接中的链路安全机制，包括根密钥使用、会话密钥生成方式，以及加密算法的选择等。消息格式为长消息编码，命令格式如表3-18所列。

表3-18 CONNECT_RSP 连接响应命令

格式	标签	长度	值域	含义
消息头	Rfu	4比特	xx	保留域，默认为0
	FormatType	4比特	8	消息格式类型
	Status	8比特	xx	响应方的当前状态
	MsgCode	8比特	18	消息码
	MsgLen	16比特	24	消息体字节长度

续表 3-18

格式	标签	长度	值域	含 义
消息体	Result	1字节	xx	Result 值定义如下： 0x00：连接成功，响应方继续等待下一步 APDU 指令或维持连接指令。 0x01：连接失败，响应方断开连接。
	RootKeyIndex	1字节	xx	根密钥索引号： ➤ 响应方确认使用的根密钥索引号，最多支持 256 组根密钥； ➤ 编码格式及定义同 CONNECT_REQ 消息相应的字段； ➤ 如果发起方选择使用动态根密钥 K0，则响应方 CONNECT_RSP 中的 RootKeyIndex 值返回 0； ➤ 如果发起方选择使用预置根密钥 Ki(i≠0)，且响应方支持发起方所选择的预置根密钥 Ki(i≠0)，则响应方 RootKeyIndex 返回密钥 Ki 对应的索引值 i； ➤ 如果发起方选择使用预置根密钥 Ki(i≠0)，但响应方不支持发起方所选择的预置根密钥 Ki(i≠0)，则响应方 RootKeyIndex 值返回 0
	SessionKey	1字节	xx	响应方确认的会话密钥生成方式： ➤ 每个比特位代表一种会话密钥生成方式，0 表示不支持该方式，1 表示支持该方式，最多支持 8 种会话密钥生成方式； ➤ 编码格式及定义同 CONNECT_REQ 消息相应的字段； ➤ 响应方在双方都支持的会话密钥生成方式中选择 1 种会话密钥生成方式，置位后返回给发起方。若双方同时支持的会话密钥生成方式有多种，则默认情况下选择其中最高位所代表的会话密钥生成方式
	EncAlg	2字节	xx	响应方确认的链路加密算法模式： ➤ 每个比特位代表一种链路加密算法模式，0 表示不支持该算法模式，1 表示支持该算法模式，最多支持 16 种加密算法模式； ➤ 编码格式及定义同 CONNECT_REQ 消息相应的字段； ➤ 响应方在双方都支持的加密算法模式中选择 1 种加密算法模式，置位后返回给发起方。若双方同时支持的加密算法模式有多种，则默认情况下选择其中最高位所代表的加密算法模式

续表 3-18

格式	标签	长度	值域	含义
消息体	SDInfo	5字节	xx	响应方信息,厂商自定义
	SDRand	8字节	xx	响应方生成的随机数,用于计算会话密钥
	Reserved	6字节	xx	保留域,默认为0
校验体	CheckSum	2字节	xx	校验值

3.5.7 APDATA_REQ 数据交换请求命令

APDATA_REQ 数据交换请求命令,发起方通过射频通道的频点 freq1(IDs)和地址 addr2(IDs)进行数据传输。用于发起方传递应用层的数据或命令,消息格式为长消息编码,命令格式如表 3-19 所列。

表 3-19 APDATA_REQ 数据交换请求命令

格式	标签	长度	值域	含义
消息头	Rfu	4比特	xx	保留域,默认为0
	FormatType	4比特	8	消息格式类型
	Status	8比特	xx	发起方的当前状态
	MsgCode	8比特	19	消息码
	MsgLen	16比特	M	消息体字节长度 $M(0 \leqslant M \leqslant 288)$
消息体	EncPayload	M 字节	xx	对明文数据(长度为 N, $0 \leqslant N \leqslant 286$)加密后得到的密文数据块,长度为 M。 ➢ 若 $(N+2)$ 为 8 的整数倍,则 $M = N+2$; ➢ 若 $(N+2)$ 不为 8 的整数倍,则 $M = [(N+2)/8] \times 8 + 8$
校验体	CheckSum	2字节	xx	校验值

3.5.8 APDATA_RSP 数据交换响应命令

APDATA_RSP 数据交换响应命令,响应方通过射频通道的频点 freq1(IDs)和地址 addr2(IDs)进行数据传输。消息格式为长消息编码,命令格式如表 3-20 所列。

表 3-20 APDATA_RSP 数据交换响应命令

格式	标签	长度	值域	含义
消息头	Rfu	4 比特	xx	保留域，默认为 0
	FormatType	4 比特	8	消息格式类型
	Status	8 比特	xx	响应方的当前状态
	MsgCode	8 比特	20	消息码
	MsgLen	16 比特	M	消息体字节长度 $M(0 \leqslant M \leqslant 288)$
消息体	EncPayload	M 字节	xx	对明文数据（长度为 N，$0 \leqslant N \leqslant 286$）加密后得到的密文数据块，长度为 M。 ➢ 若 $(N+2)$ 为 8 的整数倍，则 $M=N+2$； ➢ 若 $(N+2)$ 不为 8 的整数倍，则 $M=[(N+2)/8] \times 8+8$
校验体	CheckSum	2 字节	xx	校验值

3.5.9 LINKCTL_REQ 维持连接请求命令

LINKCTL_REQ 维持连接请求命令，发起方通过射频通道的频点 freq1(IDs) 和地址 addr2(IDs) 进行数据传输，用于在链路空闲的状态下发起方与响应方进行连接状态确认。消息格式为长消息编码，命令格式如表 3-21 所列。

表 3-21 LINKCTL_REQ 维持连接请求命令

格式	标签	长度	值域	含义
消息头	Rfu	4 比特	xx	保留域，默认为 0
	FormatType	4 比特	8	消息格式类型
	Status	8 比特	xx	发起方的当前状态
	MsgCode	8 比特	22	消息码
	MsgLen	16 比特	2	消息体字节长度
消息体	RandData	1 字节	xx	随机数
	Reserved	1 字节	xx	保留域，默认为 0
校验体	CheckSum	2 字节	xx	校验值

3.5.10 LINKCTL_RSP 维持连接响应命令

LINKCTL_RSP 维持连接响应命令，响应方通过射频通道的频点 freq1(IDs) 和地址 addr2(IDs)进行数据传输,消息格式为长消息编码,命令格式如表 3-22 所列。

表 3-22 LINKCTL_RSP 维持连接响应

格式	标签	长度	值域	含义
消息头	Rfu	4 比特	xx	保留域,默认为 0
	FormatType	4 比特	8	消息格式类型
	Status	8 比特	xx	发起方的当前状态
	MsgCode	8 比特	23	消息码
	MsgLen	16 比特	2	消息体字节长度
消息体	RandData	1 字节	xx	随机数
	Reserved	1 字节	xx	保留域,默认为 0
校验体	CheckSum	2 字节	xx	校验值

3.5.11 CHECK1_REQ 冲突检测请求命令

CHECK1_REQ 冲突检测请求命令,发送方通过磁通道发送给响应方的冲突检测命令。消息格式为短消息编码,命令格式如表 3-23 所列。

表 3-23 CHECK1_REQ 冲突检测请求命令

格式	标签	长度	值域	含义
消息头	MsgCode	4 比特	2	消息码
	MsgLen	4 比特	2	消息体字节长度
消息体	CDC	2 字节	xx	冲突检测码 CDC:取当前响应方生成的随机数 IDs 的前 2 字节。在冲突检测打开的情况下,CDC 同时作为 TRI 使用。 响应方接收到 CHECK1_REQ 后,应当按照下面的方式更新自己的当前状态: ➤ 如果 CDC/TRI 与本地 IDs 的前两字节相同,则响应方认为当前连接状态为正常(0x00); ➤ 否则响应方认为当前连接状态为异常(0x01)

3.5.12 CHECK_RSP 冲突检测响应命令

CHECK_RSP 冲突检测响应命令,响应方通过射频通道的频点 freq2(CDC)和地址 addr1(CDC)进行数据传输。消息格式为长消息编码,命令格式如表 3-24 所列。

表 3-24 CHECK_RSP 冲突检测响应命令

格式	标签	长度	值域	含义
消息头	Rfu	4比特	xx	保留域,默认为 0
	FormatType	4比特	8	消息格式类型
	Status	8比特	xx	响应方的当前状态
	MsgCode	8比特	24	消息码
	MsgLen	16比特	16	消息体字节长度
消息体	RandData	1字节	xx	随机数
	TargetID	8字节	xx	响应方标识
	Reserved	7字节	xx	保留域,默认为 0
校验体	CheckSum	2字节	xx	校验值

3.5.13 CHECK2_REQ 连接确认请求命令

CHECK2_REQ 连接确认请求命令,发起方通过磁通道发送给响应方的连接确认消息。如果当前冲突检测配置为关闭,则发起方在建立连接之后持续发送 CHECK2_REQ 连接确认消息,消息格式为短消息编码,命令格式如表 3-25 所列。

表 3-25 CHECK2_REQ 连接确认请求命令

格式	标签	长度	值域	含义
消息头	MsgCode	4比特	3	消息码
	MsgLen	4比特	2	消息体字节长度
消息体	TRI	2字节	xx	响应方随机标识码 TRI,取当前响应方生成的随机数 IDs 的前 2 字节。 响应方接收到 CHECK2_REQ 后,应当按照下面的方式更新自己的当前状态: ➢ 如果 TRI 与本地 IDs 的前两字节相同,则响应方认为当前连接状态为正常(0x00); ➢ 否则响应方认为当前连接状态为异常(0x01)

3.5.14 LTW 响应方要求等待命令

LTW 响应方要求等待命令,响应方通过射频通道的频点 freq1(IDs)和地址 addr2(IDs)进行数据传输,用于发送方申请长时间交易等待,消息格式为长消息编码,命令格式如表 3-26 所列。

表 3-26 LTW 响应方要求等待命令

格 式	标 签	长 度	值 域	含 义
消息头	Rfu	4 比特	xx	保留域,默认为 0
	FormatType	4 比特	8	消息格式类型
	Status	8 比特	xx	发起方的当前状态
	MsgCode	8 比特	25	消息码
	MsgLen	16 比特	2	消息体字节长度
消息体	RandData	1 字节	xx	随机数
	Reserved	1 字节	xx	保留域,默认为 0
校验体	CheckSum	2 字节	xx	校验值

3.5.15 CLOSE_REQ 关闭连接请求命令

CLOSE_REQ 关闭连接请求命令,发起方通过射频通道的频点 freq1(IDs) 和地址 addr2(IDs) 进行数据传输,用于发起方断开与响应方的连接,消息格式为长消息编码,命令格式如表 3-27 所列。

表 3-27 CLOSE_REQ 关闭连接请求命令

格 式	标 签	长 度	值 域	含 义
消息头	Rfu	4 比特	xx	保留域,默认为 0
	FormatType	4 比特	8	消息格式类型
	Status	8 比特	xx	发起方的当前状态
	MsgCode	8 比特	26	消息码
	MsgLen	16 比特	4	消息体字节长度
消息体	NeedResp	1 字节	0/1	是否需要返回关闭连接确认(默认为 0)。 ➢ 0:无需 CLOSE_RSP 响应; ➢ 1:需要 CLOSE_RSP 响应
	Reserved	3 字节	xx	保留域,默认为 0
校验体	CheckSum	2 字节	xx	校验值

3.5.16 CLOSE_RSP 关闭连接响应命令

CLOSE_RSP 关闭连接响应命令，响应方通过射频通道的频点 freq1(IDs) 和地址 addrs(IDs) 进行数据传输，响应方对 CLOSE_REQ 的响应确认，只有当发起方在 CLOSE_REQ 中指明需要响应方对关闭连接进行确认（NeedResp＝1）时，响应方才做 CLOSE_RSP 响应确认，否则响应方在关闭连接后立即结束，不做 CLOSE_RSP 响应。消息格式为长消息编码，命令格式如表 3－28 所列。

表 3－28 CLOSE_RSP 关闭连接响应命令

格式	标签	长度	值域	含义
消息头	Rfu	4 比特	xx	保留域，默认为 0
	FormatType	4 比特	8	消息格式类型
	Status	8 比特	xx	发起方的当前状态
	MsgCode	8 比特	27	消息码
	MsgLen	16 比特	4	消息体字节长度
消息体	CloseResult	1 字节	0/1	关闭结果： ➢ 0：响应方关闭成功； ➢ 1：响应方关闭失败。
	Reserved	3 字节	xx	保留域，默认为 0
校验体	CheckSum	2 字节	xx	校验值

3.6 RCC 限域通信技术与主流近场支付方案的对比

目前，作为非接触交易的两种移动支付线下应用技术 RCC 和 NFC 的最大差别是：前者因为频段的不同需要去改造 POS 机读头部分，另外在用户端需更换一张支持 RCC 的 SIM 卡；后者虽然可以沿用传统的 13.56 MHz 的非接触环境，但是需要在用户的支付终端上进行硬件和软件的集成才能支持支付功能。表 3－29 把 RCC 与 NFC 技术做了一个全方位的对比。

注意：表 3－29 中表述的强、弱、快、慢、大、小等的比较，是针对 RCC 和 NFC 技术之间的比较，另外在支持支付终端上，RCC 技术方案在多应用支持方面相比 NFC

技术还有提升的空间。

表 3-29　RCC 限域通信技术与主流近场支付方案对比

类　型	RCC	NFC
安全性	强	强
交易速度	快	快
通信方式	13.56 MHz 射频通道	2.45 GHz 射频通道＋磁通道
受理环境改造	大	小
支付终端改造	小	大
知识产品分布	中国企业拥有核心专利	外国企业拥有核心专利
空中发卡支持	强	强
多应用支持	弱	强
市场认知度	低	高
商用案例	深圳、哈尔滨、青岛公交等	Apple Pay、Huawei Pay、Mi Pay 等

NFC 技术不管是采用 eSE 方案还是采用 SWP SIM 方案，都可以支持网络发卡方式，通过 SE 安全模块的接触通道对卡片应用进行发卡和充值，查询余额等业务也可以通过 SE 接触通道完成，这样用户就不必跑到专门的柜台去办理相关业务，而且在 NFC 技术中，例如 Huawei Pay 和 Mi Pay 在 SE 安全模块里已经支持了多卡应用和多种不同业务的管理，现已商用并支持多张银行卡以及多张公交卡片共存。

4 MST 磁安全传输支付系统

磁安全传输（MST）是一种通过发出的磁信号模拟传统的磁条卡的技术。其应用目标为把现有磁条卡的一些业务，除现有大量市场存量的银行磁条卡外，还有积分卡、优惠券、会员卡等集成到支持 MST 的智能设备中。MST 磁信号从用户终端设备发送到支付终端的读卡器，最终实现模仿刷物理磁条卡的过程，这个支付流程无需升级改造支付读卡终端的软件或硬件。MST 技术几乎支持所有支付终端和磁条读卡器，可能有个别传统的磁条支付终端需要进行系统软件更新。

目前，韩国三星公司是这个技术的主要引导者，其旗下出品的支持 Samsung Pay 三星智付的智能手机大部分都支持 MST 技术。其实 MST 技术本身也是基于传统的磁条卡技术的，与它一脉相承，所以本章会有部分篇幅来介绍传统磁条卡业务的技术原理等。该技术自 2012 年由 Will Graylin 和 George Wallner 组建的 LoopPay 公司发明推广后，再到三星公司集成该技术到其手机中进行商用，其间经历的重要时间节点如下：

2012 年，Will Graylin 和 George Wallner 共同组建 LoopPay 公司，并且向外正式推广其发明的 MST 技术；

2013 年 11 月，贝塔斯曼旗下专注于高新科技以及生物科技的早期天使基金 Beta Fund，对 LoopPay 投资 1 000 万美元；

2014 年 7 月，Visa 联合同步金融 Synchrony Financial（前身为通用电气 GE 消费金融业务部）投资 LoopPay 公司；

2015 年 2 月，LoopPay 公司被韩国三星公司收购，三星公司计划在其之后发布的移动支付的手机中集成 MST 的技术；

2016 年 3 月，三星公司联合银联正式在中国大陆推出其移动支付产品 Samsung Pay，并且在其手机 Galaxy S7、Galaxy S7 Edge、Galaxy Note 7 等及后面发布的手机中，陆续支持 MST 技术。

MST 技术的工作过程：当用户在钱包端通过特殊的磁条读卡器或手动输入原始的物理磁条卡片信息后，就相当于把相关卡片进行了一个从磁条到手机的绑定程序,当要进行支付时,通过快捷方式调出和选择好相关卡片就可以进行支付了。支付过程其实也是通过一个非接触的方式进行的,当把激活好相关卡片的手机靠近传统的磁条刷卡机时就可以完成支付了。三星手机目前发送支付信息的有效距离为手机与磁条支付终端的磁中心点之间小于 3 cm 的区域。磁场过程中传递的交易和支付信息将会通过保密和一次性的标记号码的技术(有的资料也把它称为令牌技术)保证通信安全,总体上来讲 MST 技术会比使用传统磁条卡进行支付更安全。

通过上述简单的交易过程描述可知,目前使用的 MST 技术不只是简单地把传统磁条卡的信息通过磁场模拟的方式发送出去,而且在传统的磁条卡技术的基础上增加了安全等级,例如使用了支付标记化(Payment Tokenization)的技术。在国内外主流的一些标准化组织发布的支付标记化技术框架中,在充分考虑到互操作性、兼容性的基础上,解决了卡号信息泄露、支付场景认证等方面的问题。与传统的磁条卡的非支付标记化方案不同,支付标记在确保与现有的支付流程进行融合的同时,有效加强了身份认证与风险监控,会由信赖的标记服务商根据主账号(PAN)生成一个替代码,一般由 13～19 位的数字组成,该数值符合主账号的基本验证规则,当然也包括 Luhn 算法校验。在银行卡支付交易中用支付标记替换卡号,用支付标记的有效期替换卡号有效期,这样既不会影响交易处理,又增强了交易安全。所以对于 MST 所实现的支付标记化技术而言,其工作流程如图 4-1 所示,当用户选取好卡片且通过了个人身份验证后准备进行支付时,手机会先把这些之前绑卡到手机的一些关联数据通过加密方式传递给账号运营管理服务器,这个服务器负责进行身份

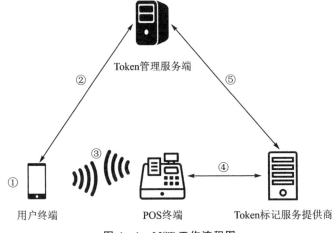

图 4-1　MST 工作流程图

验证和鉴权，通过后则会通过信赖的标记提供商下发一次性的标记号码给手机，手机收到标记号码后会把它通过磁场方式传递给磁条POS终端，再之后POS终端与标记后台和发卡行之间就会运用标准的支付标记化技术完成支付。

因为MST技术本身并不是针对一个高安全等级的存储技术，所以在收到一次性标记号码时，或者网络联网质量有问题时，需要生成一个临时标记号用于支付。一般现在主流的方法还是会借助底层的TEE或者eSE技术，在一个安全环境下通过TA或者Applet应用来生成这个号码，以达到端到端的安全风险控制。当然其前提条件是在TEE或者eSE中开发和植入一个具备算法能力的安全应用程序，之后在需要离线生成标记号时，就可以在用户的终端系统中闭环完成这个部分。

1. 用户终端能正常联网情况下的技术过程分析

① 用户终端网络数据正常连接状态，当用户开启支付应用程序时，要求用户进行个人身份验证，例如使用指纹、虹膜或者人脸识别等技术，整个个人身份验证的过程会在TEE中完成；验证通过后系统会通过网络加密传输技术把即将要支付的卡数据的主账号（PAN）以及设备唯一关联号，传递给Token标记管理服务端，转入步骤②。

② Token标记管理服务端，在接收到卡数据的主账号（PAN）以及设备唯一关联号后，会经历一个卡数据的验证过程，准确无误后，会重新签发一个一次性的Token标记号给用户终端，用户终端收到后转入步骤③。

③ 用户把支付终端靠近POS终端的磁头3 cm附近，POS终端接收到磁信号后解密出交易类型和用于使用一次性交易的Token标记号，POS终端做一个基本数据逻辑确认，没有问题后把数据通过网络加密的方式上传给Token标记服务提供商，转入步骤④。

④ Token标记服务提供商从POS终端上收到交易订单号和一次性的Token标记号码后，会做Token标记号码类型确认，如果属于静态Token标记号码则比对数据，无误则核准这笔交易，并且通过图4-1中的通道④告知POS终端此笔交易有效；如果Token标记号码类型为动态，且为一次性的交易标记号码，则Token标记服务提供商会和Token标记管理服务端进行数据加密通信，交换和比对一次性交易的Token标记号，也就是步骤⑤。

⑤ Token标记服务提供商因为收到了一次性的Token标记号码，但自己的数据库中并没有这笔数据，所以就需要和Token标记管理服务端进行数据同步，如果比

对确定了在一个定义的时间段内一次性的 Token 标记号码还处于存活状态,则认可这笔交易有效,其也是通过通道④告知 POS 终端此笔交易有效,交易成功并关闭;假如一次性的 Token 标记号码有异常或者已经超出了定义的数据存活时间,则判定这笔交易无效,通过通道④告知 POS 终端此笔交易无效,交易失败并关闭。

其实对于实际用户来讲并不会太关心底层的具体接口技术,例如目前使用的 MST 支付技术系统,在底层技术方面实际上相当于包含了 NFC 技术和 MST 技术。上面介绍的情况为用户的支付终端在联网下的支付过程,那么在一些无法连接网络或者网络通信质量不佳的情况下是否可以进行交易支付呢?答案是可以的,前提条件为应用程序在前次成功联网的情况下通过 TEE 往 eSE 安全芯片里存放一个特殊的标记码,用于这种情况下的支付。

2. 无法连接网络或者网络通信质量不佳情况下的技术过程分析

MST 典型工作框图如图 4-2 所示。

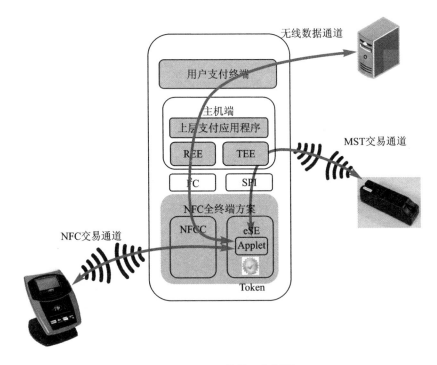

图 4-2 MST 典型工作框图

① 当用户在手机上打开上层支付应用程序时,在电量以及网络通信质量正常的情况下,应用程序会尝试在后台把手机端的 eSE 唯一特征值(通常情况下为 SE 的

CPLC 数据,有些地方也会把 CPLC 数据做拆解和重新分散出来一个唯一的 SEID 数据)送给 TSM 服务器端,TSM 服务器端收到 eSE 的唯一特征值以及业务请求代码,例如这里的业务请求就是希望下发管理 MST 的 Token 标记号的 Applet 应用程序,那么这时 TSM 服务端就会把 eSE 的唯一特征值送到 HSM 硬件加密机里;HSM 硬件加密机会根据之前部署好的分散算法以及分散因子,把 eSE 的安全域的密钥送回到 TSM 端;TSM 端拿到 eSE 的主控安全域或者辅助安全域(视具体的业务和方案而定)的密钥后,就会对准备好即将要下发的管理 MST 的 Token 标记号的 Applet 应用程序进行分包和加密,使之成为数据脚本格式,然后通过网络加密通信的方式把数据脚本送回给主机端;上层支付应用程序就会把数据通过 TEE 的 TA 程序经 SPI 总线透传给 eSE 安全单元,当 eSE 单元收到正确的密钥后,就会建立起一个端到端的 SCP 安全通道,把 TSM 服务器端下发的数据脚本安装到 eSE 里。这个全过程就是一个提前下载安装管理 MST 的 Token 标记号的 Applet 应用程序的过程。

② 当用户在支付端通过特殊磁头读取或者手动输入卡信息进行绑卡后,上层支付应用程序把卡信息和设备唯一关联号码通过网络 SSL 协议加密的方式与后端服务器之间进行通信。

③ 当服务器后端认为卡信息有效时,TSP 标记服务方会把卡数据的主账号(PAN)以及设备唯一关联号,进行安全存储并会签发一个 Token 标记号码送回给主机端,上层支付应用程序收到后,会通过 TEE 中的 TA 再经 SPI 总线把 Token 标记号给 eSE 中管理 Token 标记号的应用程序。

④ 对于极端情况,例如无法连接网络或者网络通信质量不佳,用户设备在准备进行 MST 技术支付时,就可以不用联网的方式去获得一次性的 Token 标记号进行支付,而是可以离线从本机 eSE 中取出 Token 标记号,通过 MST 技术发送出去用于支付。

⑤ 为防止用于支付用途的 Token 标记号出现安全风险,当用户的支付设备终端再次联网时,在后台以及 eSE 中会动态再重新生成及存储一次 Token 标记号,注意为安全起见,这个步骤可以是定期或者非定期动态双方更新的过程,视具体的业务和环境而定。

Luhn 算法

Luhn 算法或者 Luhn 公式,也被称作"模 10 算法"。它是一种简单的校验公式,一般会被用于身份证号、手机 IMEI 号和银行卡号等领域,计算证件号的合法性。该算法是由IBM公司的科学家Hans Peter Luhn所创造的,其设计之初的目的并不是为了使之成为一种类似数据加密安全的哈希函数,而是为了防止一些意外出现的数据错误,其主要目的是使之能方便并广泛地使用在一些算法计算相对要求不高的主机端或者机器设备上,能快速计算出校验码,这样就能快速判断出证件号的数据逻辑合法性,使算法设计更精简和高效。

Luhn算法示例过程步骤如下:

① 从证件号最后一位数字开始,偶数位乘以2,如果乘以2的结果是两位数,将结果减去9。

② 把所有数字相加,得到总和。

③ 如果证件号是合法的,总和可以被10整除。

例如一个证件号为"5432123456788881",则奇数位分布为42246881,偶数位分布为53135788。如图4-3所示为数据的Luhn求值过程,奇数位求和的结果为$4+2+2+4+6+8+8+1=35$;偶数位乘以2,结果大于或等于10的要减去9,偶数位求和的结果为$(5\times2-9)+(3\times2)+(1\times2)+(3\times2)+(5\times2-9)+(7\times2-9)+(8\times2-9)+(8\times2-9)=35$。最后$35+35=70$可以被10整除,认定校验通过。

号码	5	4	3	2	1	2	3	4	5	6	7	8	8	8	8	1
偶数位	1		6		2		6		1		5		7		7	
奇数位		4		2		2		4		6		8		8		1
奇偶数总和	70															

图4-3 Luhn求值示例

下面分别为Luhn算法的面向机器、对象以及脚本语言的参考代码,其中C语言代码示例如下:

```
#include <stdlib.h>  // atoi
#include <string.h>  // strlen
#include <stdbool.h> // bool

bool checkLuhn(const char * pPurported)
{
    int nSum = 0;
    int nDigits = strlen(pPurported);
    int nParity = (nDigits - 1) % 2;
    char cDigit[2] = "\0";
```

```
        for (int i = nDigits; i > 0 ; i--)
        {
            cDigit[0] = pPurported[i - 1];
            int nDigit = atoi(cDigit);

            if (nParity == i % 2)
            nDigit = nDigit * 2;

            nSum += nDigit/10;
            nSum += nDigit % 10;
        }
        return 0 == nSum % 10;
}
```

JAVA 语言代码示例如下：

```
public static boolean check(int[] digits) {
    int sum = 0;
    int length = digits.length;
    for (int i = 0; i < length; i++) {

        // get digits in reverse order
        int digit = digits[length - i - 1];

        // every 2nd number multiply with 2
        if (i % 2 == 1) {
            digit *= 2;
        }
        sum += digit > 9 ? digit - 9 : digit;
    }
    return sum % 10 == 0;
}
```

Python 语言代码示例如下：

```
def checkLuhn(purportedCC = ''):
    sum = 0
    parity = len(purportedCC) % 2
    for i, digit in enumerate([int(x) for x in purportedCC]):
        if i % 2 == parity:
```

```
        digit *= 2
        if digit > 9:
            digit -= 9
    sum += digit
return sum % 10 == 0
```

4.1 磁条卡技术原理

磁条有时被称为磁卡或超级条码，它是一种卡片状的磁性存储数据的介质，通过修改微小的磁性铁基材料的磁性粒子来记录字符与数字信息。常用的磁条卡有银行发行的借记卡或者信用卡、会员卡、公交卡、校园卡、电话卡、电子游戏卡、积分卡等。

在 20 世纪 70 年代之前，银行卡交易更多的是通过物理方式进行，而非数字形式，每笔交易都采用一个微小的印刷机将卡上的凸起字母和数字拓印到双面压敏纸上进行记录，然后每张纸被传送到处理中心，由人工录入到计算机系统中。这种方式虽然在某些无法连接网络的特殊场合还在使用（例如高空或者远海），但是对于绝大部分的场景，由于其系统安全性差、速度缓慢、容易出错，而被磁条卡所取代。磁条卡由强度高、耐高温的塑料或纸质涂覆塑料制成，防潮、耐磨并且有一定的柔韧性，携带方便，使用较为稳定可靠，造价便宜，在记录交易时只需轻轻一刷就可以快速完成，因其给人们带来的便捷性，使当前全世界每年通过"磁条"读卡器刷卡的人超过 500 亿次。

丹麦于 1900 年左右发明了使用磁记录钢卷尺和线记录音频信息，1960 年 IBM 公司在与美国政府合作安全保障体系时，利用磁带技术原理开发了一个基于磁条卡技术的全新方式。1969 年 IBM 公司的工程师福勒斯特·帕里（Forrest Parry）尝试使用各种类型的胶去粘合磁带和塑料卡片，但每次磁带扭曲问题或其胶粘剂处呈现磁带无法正常读取使用的问题使他一度非常沮丧。有一天，他像往常一样在实验室工作了一天后还是没有得到实际的进展，回到家后，他向正在熨烫衣服的妻子多萝西娅（Dorothea）解释他最近沮丧的原因：不能让磁带"贴"到塑料上进行工作。妻子建议他用熨斗融化磁带试试。后来福勒斯特·帕里采用了这个建议，成功地实现了磁条粘合。这也就有了现在人们大量使用的磁条卡片。

4 MST 磁安全传输支付系统

4.1.1 规范协议族

国际标准 ISO/IEC 7810、ISO/IEC 7811、ISO/IEC 7812、ISO/IEC 7813、ISO 8583、ISO/IEC 15457 和 ISO/IEC 4909 对磁条卡的物理属性进行了定义,包括大小、灵活性、超级条码的位置、磁特性、数据格式等,此外还包含了金融标准卡片,如信用卡号码的分配范围和定义不同的信用卡发行机构等。磁条卡片标准的定义和范围如表 4-1 所列。

表 4-1 标准的定义和范围

标准号	标准名称	发布日期
ISO/IEC 7810	识别卡-物理特性	2003-11-01
ISO 7811-1—1985	识别卡-录制技术 第1部分:压印	1985-12-01
ISO 7811-1—1995	识别卡-记录技术 第1部分:压印	1995-08-15
ISO/IEC 7811-1—2002	识别卡-记录技术 第1部分:凸印	2002-09-15
ISO/IEC 7811-1—2014	识别卡-记录技术 第1部分:凹凸印	2014-08-25
ISO 7811-2—1985	识别卡-录制技术 第2部分:磁卡	1985-12-01
ISO 7811-2—1995	识别卡-录制技术 第2部分:磁卡	1995-08-15
ISO/IEC 7811-2—2001	识别卡-录制技术 第2部分:磁条-低矫顽磁力	2001-02-01
ISO/IEC 7811-2—2014	识别卡-记录技术 第2部分:磁条-低矫顽磁力	2014-08-14
ISO 7811-3—1985	识别卡-录制技术 第3部分:ID-1 卡片压印字符的位置	1985-12-01
ISO 7811-3—1995	识别卡-记录技术 第3部分:在 ID-1 卡片压印字符的位置 第2版	1995-08-15
ISO/IEC 7811-3—1995	识别卡-记录技术 第3部分:ID-1 卡片压印字符的位置	1995-08-01
ISO 7811-4—1985	识别卡-录制技术 第4部分:只读磁道的位置-单列磁道和双列磁道	1985-12-01
ISO 7811-4—1995	识别卡-录制技术 第4部分:只读磁道的位置-单列磁道和双列磁道	1995-08-15
ISO 7811-5—1985	识别卡-录制技术 第5部分:读写磁道位置-三列磁道	1985-12-01
ISO 7811-5—1995	识别卡-录制技术 第5部分:读写磁道位置-三列磁道	1995-08-15
ISO 7811-6—1996	识别卡-录制技术 第6部分:磁卡-高矫顽磁力	1996-04-15
ISO/IEC 7811-6 AMD 1—2005	识别卡-记录技术 第6部分:磁卡-高矫顽磁力 修改件1标准和试验方法	2005-10-01

续表 4-1

标准号	标准名称	发布日期
ISO/IEC7811-7	识别卡-录制技术 第7部分:磁卡-高矫顽磁力,高密度	2004-07
ISO/IEC7811-8	识别卡-录制技术 第8部分:磁卡-矫顽磁力 51.7 kA/m (650 Oe)	2008-03
ISO/IEC 7811-9	识别卡-录制技术 第9部分:磁卡-触觉识别标识	2008-06
ISO/IEC 7812-1	识别卡-发行者识别 第1部分:编号系统	2017-01
ISO/IEC 7812-2	识别卡-发行者识别 第2部分:应用和注册程序	2017-01
ISO/IEC 7813—2006	信息技术-识别卡-金融交易卡	2006-07
ISO/IEC 17811-1—2014	信息技术-设备控制和管理 第1部分:体系结构	2014-06-05
ISO/IEC 17811-2—2015	信息技术-设备控制和管理 第2部分:设备控制和管理协议规范	2015-02-26
ISO/IEC 17811-3—2014	信息技术-设备控制和管理 第3部分:可靠消息传递协议规范	2014-09-23
ISO 8583-1—2003	金融交易卡原始信息-交换消息规范 第1部分:消息、数据元素和码值	2003-06
ISO 8583-2—1998	金融交易卡原始信息-交换消息规范 第2部分:应用程序和登记程序机构标识码(IIC)	1998-06
ISO 8583-3—2003	金融交易卡原始信息-交换消息规范 第3部分:维护程序信息、数据元素和代码值	2003-05
ISO/IEC 4909—2006	识别卡-金融交易卡-磁条数据内容三磁道	2006-07
ISO/IEC 15457-1—2008	识别卡-柔性薄卡 第1部分:物理特性	2008-03
ISO/IEC 15457-2—2008	识别卡-柔性薄卡 第2部分:磁记录技术	2007-06
ISO/IEC 15457-3—2008	识别卡-柔性薄卡 第3部分:测试方法	2008-03

近些年国内有些银行新发的银行卡可以支持磁条和芯片两种工作方式。这种类型的银行卡在与POS端交易时可以支持三个通道,即可以通过接触磁条、接触芯片和非接触芯片进行交易,这种类型的卡也称为三界面卡;还有的卡不支持非接触芯片的方式,只支持两个通道,称为双界面卡。对于这种双界面或者三界面的复合卡的使用指导,请参照央行于2016年下发的《关于进一步加强银行卡风险管理的通知》,银行于2017年5月1日起全面关闭芯片磁条复合卡的磁条交易。实际上芯片磁条复合卡是金融IC卡的过渡卡种,可同时支持芯片和磁条两种介质。目前在具备芯片受理能力的受理终端上,推荐使用通过芯片发起支付交易;银行卡芯片化迁移

过渡期间,在不具备芯片受理能力的受理终端可以通过磁条发起支付交易。

识别卡的物理特性

在标准 ISO/IEC 7810 中详细定义了平时日常生活中所能见到的各种大小卡片的尺寸,目前一共定义了四种类型,分别为 ID-1、ID-2、ID-3 和 ID-000,比如我们平时使用的银行卡不管是磁条卡还是芯片卡,还有标准的公交卡,基本都是 ID-1 格式的,表 4-2 所列为 ISO/IEC 7810 定义的各种类型卡的尺寸和用途。

表 4-2 卡的类型、尺寸和用途

类　　型	尺　　寸	主要应用方向
ID-1	85.60 mm×53.98 mm	银行卡、身份证和标准公交卡
ID-2	105 mm×74 mm	某些国家的身份证和签证
ID-3	125 mm×88 mm	护照和通行证
ID-000	25 mm×15 mm	标准的 Mini SIM

上述各种类型卡其厚度统一为 0.76 mm,图 4-4 所示为结合标准 ISO/IEC 7816 定义的卡类型示意图。

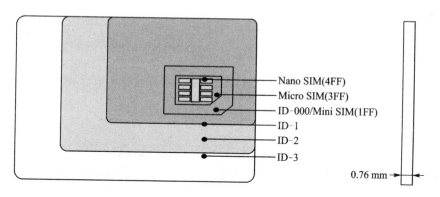

图 4-4 卡类型示意图

如图 4-4 所示 Mini SIM(1FF)、Micro SIM(3FF)和 Nano SIM(4FF)的尺寸分别为 25 mm×15 mm、15 mm×12 mm 和 12.30 mm×8.80 mm。

4.1.2 尺寸参数

根据 ISO/IEC 7811 标准中定义的磁卡通常包含 3 个磁道,分别为磁道 1、2、3 (Track1,2,3),每个磁道的宽度是相同的,用来衡量磁条抵抗因受外界磁场影响而造成数据损失的能力,称磁抗又称抗消磁性,描述磁抗(矫顽磁力)的单位为 OE(奥斯特)。磁抗的类型分为两类:低磁抗条,为具有普通抗消磁性磁条,磁抗一般为300～650 OE;高磁抗条,为具有较高抗消磁性磁条,磁抗一般为 2 750 OE、3 500 OE 和 4 000 OE。目前市面上主流磁卡的参考尺寸数据如下:

- 每个磁道在 2.80 mm(0.11 in)左右,用于存放用户的数据信息。
- 相邻两个磁道之间约有 0.5 mm(0.02 in)的间隙,方便区分相邻的两个磁道。
- 如果磁卡应用的是支持两个磁道的,那么整个磁带的宽度为 6.35 mm(0.25 in)左右,这种类型的卡在市面上并不多见,主要应用于一些特殊的行业。
- 如果磁卡应用的是支持三个磁道的,那么整个磁带的宽度为 10.29 mm(0.405 in)左右,这种卡在人们的日常生活中被大量使用,实际上我们所接触、看到的银行磁卡上的磁带宽度会加宽 1～2 mm 左右,磁带总宽度在 12～13 mm 之间。
- 一般在磁带上记录的三个有效磁道数据的起始数据位置和结束数据位置,不会在磁带的边缘,而是在磁带边缘向内约 7.44 mm(0.293 in)为起始数据位置(引导0区),在磁带边缘向内约 6.93 mm(0.273 in)为终止数据位置(尾随0区)。这些标准主要是为了有效保护磁卡上的数据不易被丢失,因为磁卡边缘上的磁记录数据比较容易因为物理磨损而被破坏。
- 磁条区域与标准卡的顶边之间最大间距为 5.54 mm。

磁卡目前在银行领域还是有海量的市场,绝大部分的磁条银行卡都是基于 ID-1 的外形尺寸。如图 4-5 所示为标准三磁道磁卡的尺寸结构。

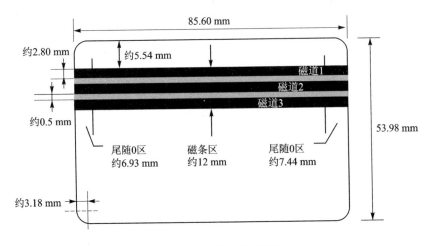

图 4-5 磁条卡尺寸示意

4 MST 磁安全传输支付系统

银行磁卡每个磁道的位置可以上下偏移 0.5 mm 左右,也就是说同一磁条边沿最上可以到达的位置与最下可以到达的位置之间的最大距离为 1 mm,每个磁道的宽度一般为 2.8 mm 左右,而一般读磁卡的磁头要求磁道宽度为 1.4 mm±0.1 mm 左右,另外磁头定位孔允许偏差的距离在±0.05 mm 范围之内。在最极限的情况下,普通磁头的定位偏差可能有 0.1 mm 左右,这样磁头磁道宽度就在 1.5 mm 左右。一般磁头磁条定义的允许活动空间在 1.7 mm 左右,因此只要所有构件定位均符合设计要求,则无论偏上边沿刷卡还是偏下边沿刷卡,在进行刷卡交易时,磁头磁条始终保持在卡片磁道范围内,从而保证了刷卡的兼容性和可靠性。

通常一个磁头刷卡器的设计需要注意卡槽、磁头弹片、弹片固定盖、读卡器材料等,本节不会对具体的安装性的参数和材料做过多介绍,而是重点介绍磁头读卡器的尺寸和几个设计注意事项。如图 4-6 所示为普通读磁卡磁头的参考尺寸数据。

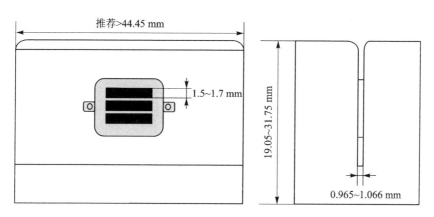

图 4-6 磁头尺寸示意

下面介绍磁头读卡器设计过程中需要注意的事项:

➢ 磁头中心线与固定孔之间的中心线需要在同一条线上,卡片磁道 2 的中心线与磁头第二轨的中心线要在同一条线上。

➢ 磁头卡槽部分为了更好地兼容 0.254～0.838 mm 的卡,在设计读头卡槽宽度时基本上会推荐在 0.864～0.965 mm 之间,卡槽的宽度精度需要控制在 0.101 mm 以内,也就是说卡槽的最终宽度推荐控制在 0.965(0.864 mm + 0.101 mm)～1.066(0.965 mm + 0.101 mm) mm 之间。

➢ 磁头卡槽的高度推荐在 19.05～31.75 mm 之间,下限设置在 19.05 mm 是为了避免刷卡片时过度倾斜,上限设置在 31.75 mm 是为了避免卡片的压纹字在刷卡时进入到卡槽中。
- 刷卡槽的整个长度设计推荐大于 44.45 mm,如果距离过短,那么在刷卡过程中可能会使卡片在进入卡槽时往下倾斜,造成卡片磁道偏离磁头,导致读卡成功率降低。
- 在磁卡读头上推荐加入一个固定弹片,以方便刷卡。

4.1.3 磁道数据详解

如上所述磁条上会有三个磁道。一般情况下磁道 1 与磁道 2 的数据为只读属性,即在使用时,磁道上记录的信息只能读出而不允许改写或者修改;磁道 3 的数据为读/写属性,即在使用时既可以读出,也可以写入。如表 4-3 所列为三个磁道的信息以及格式属性。

表 4-3 三磁道数据

磁道号	间距	规范	密度	数据编码	内容
无磁区	5.54 mm		N/A		
磁道 1	2.8 mm	IATA	827 bits/mm (210 bpi)	7 比特 6 数据比特+1 奇偶校验位①	最多支持 79 个字母数字字符
磁道 2	2.8 mm	ABA	295 bits/mm (75 bpi)	5 比特 4 数据比特+1 奇偶校验位	最多支持 40 个数字字符
磁道 3	2.8 mm	THRIFT	827 bits/mm (210 bpi)	5 比特 4 数据比特+1 奇偶校验位	最多支持 40 个数字字符
无磁区			N/A		

① 奇偶校验位:一般在无特别说明的情况下,磁卡采用奇校验,当接收端收到这组代码时,数据体和校验位一起校验"1"的个数,为奇数时正确。在算法应用时也非常简单,把数据体的二进制值进行相加,如果为奇数,则对校验位填 0;反之如果为偶数(包含全为零),则对校验位填 1。

1. 磁道 1 的信息以及格式属性

- 本磁道的数据格式属性等由国际航空运输协会(IATA,International Air Transport Association)定义。
- 记录密度为 827 bits/mm(210 bpi)。
- 可以记录数字(0~9)、字母(A~Z)和其他一些特殊符号(如括号、起始符、

4 MST 磁安全传输支付系统

终止符和分隔符等)。
- 最多可记录 79 个数字或字母符号。
- 每个数据(一字节)由 7 比特组成。
- 本磁道上的信息一般记录了磁卡的使用类型、范围等一些标记和说明性的信息,例如银行用卡系统中,会记录用户的姓名、卡的有效使用期限以及其他的一些特殊"标记"信息。

磁道 1 的数据格式如图 4-7 所示。

SS	FC	76 个数字或字母符号					ES	LRC
		数据		数据		数据		
		PAN	FS	NAME	FS	Additional Data	Discretionary Data		
磁道 1 数据示例								
%	B	1234567890123445	^	XIAOHUA WANG	^	2010120	00000000000000＊＊xxx＊＊＊＊＊＊	?	＊

图 4-7 磁道 1 的数据格式

SS:Start Sentinel	%	0x05	起始符
FC:Format Code	N/A	N/A	格式码(1 字符)
FS:Field Seperator	^	0x3E	分隔符
ES:End Sentinel	?	0x1F	终止符
LRC:Longitudinal Redunancy Check Characte	N/A	N/A	纵向冗余校验位

PAN(Primary Account Number,主账号):用于标识发卡机构及卡片的号码,由发卡机构标识代码、发卡机构自定义位和校验位组成。通常情况下不能超过 19 个数字。

NAME(账号名):主要用于记录与账号相关联的用户名字或者标识。通常情况下最多由 26 个字符或者数字字符组成。

Additional Data(额外数据):此域在银行卡系统中,主要数据包含 4 个字符码用来记录卡片的截止日期,格式为 YYMM,例如 2017 年 10 月,则编码为 1710;3 个字符码用来记录服务代码。

Discretionary Data(授权数据):此域在银行卡系统中,主要数据包含 1 个字符码用来记录 PIN 校验密钥索引码(PVKI,PIN Verification Key Indicator);4 个字符码用来记录 PIN 校验值(PVV, PIN Verification Value);3 个字符码用来记录卡校验值(CVV, Card Validatoin Value)或者卡校验码(CVC, Card Validation Code)。

2. 磁道 2 的信息以及格式属性

- 本磁道的数据格式属性等由美国银行家协会(ABA, American Bankers Association)定义。
- 记录密度为 295 bits/mm (75 bpi)。
- 可以记录数字(0~9)和一些特殊符号(如起始符、终止符、分隔符和控制位等),不记录字母等信息。
- 最多可记录 40 个数字符号。
- 每个数据(一字节)由 5 比特组成。
- 本磁道上的信息一般记录了磁卡用户的账户信息、款项信息等,当然还有一些银行所要求的特殊信息等。

磁道 2 的数据格式如图 4-8 所示。

SS	37 个数字或字母符号				ES	LRC
	数据	FS	数据			
	PAN		Additional Data	Discretionary Data		
磁道 2 数据示例						
;	1234567890123445	=	2010120	0xxxx00000000	?	*

图 4-8 磁道 2 的数据格式

SS:Start Sentinel   ;   0x0B   起始符
FS:Field Seperator   =   0x0D   分隔符
ES:End Sentinel   ?   0x0F   终止符
LRC:Longitudinal Redunancy Check Characte   N/A   N/A   纵向冗余校验位

PAN(Primary Account Number,主账号):用于标识发卡机构及卡片的号码,由发卡机构标识代码、发卡机构自定义位和校验位组成。通常情况下不能超过 19 个数字。

Additional Data(额外数据):此域在银行卡系统中,主要数据包含 4 个字符码用来记录卡片的截止日期,格式为 YYMM,例如 2017 年 10 月,则编码为 1710;3 个字符码用来记录服务代码。

Discretionary Data(授权数据):此域在银行卡系统中,主要数据包含1个字符码用来记录 PIN 校验密钥索引码(PVKI);4个字符码用来记录 PIN 校验值(PVV);3个字符码用来记录卡校验值(CVV)或者卡校验码(CVC)。

3. 磁道3的信息以及格式属性

- 本磁道的数据格式属性等由定期存款行业(THRIFT,Thrift‑Savings Industry)定义。
- 记录密度为 827 bits/mm (210 bpi)。
- 可以记录数字(0~9)和一些特殊符号(如起始符、终止符、分隔符和控制位等),不记录字母等信息。
- 最多可记录 107 个数字符号。
- 每个数据(一字节)由 5 比特组成。
- 本磁道上的信息一般记录了磁卡用户的账户信息、款项信息等,当然还有一些银行所要求的特殊信息等。

磁道3的数据格式如图 4‑9 所示。

SS	FC	数据	FS	104 个数字或字母符号		ES	LRC
				数据			
		PAN		User and Security Data	Additional Data		
磁道3数据示例							
;	01	1234567890123445	=	724724100000000000030300xxxx040400099010	0000000000000000	?	*

图 4‑9 磁道3的数据格式

SS:Start Sentinel　　　　　　　　　　　　　;　　0x0B　起始符
FC:Format Code　　　　　　　　　　N/A　N/A　格式码(2字符)
FS:Field Seperator　　　　　　　　　　=　　0x0D　分隔符
ES:End Sentinel　　　　　　　　　　　?　　0x0F　终止符
LRC:Longitudinal Redunancy Check Characte　N/A　N/A　纵向冗余校验位

PAN(Primary Account Number,主账号):用于标识发卡机构及卡片的号码,由发卡机构标识代码、发卡机构自定义位和校验位组成。通常情况下不能超过 19 个数字。

User and Security Data(用户及安全数据):根据 ISO/IEC 4909 的定义,此域主

要包括如下数据,准确的数据还需要与对应银行卡发卡机构进行确认。

- 3 个字符码记录可选项国家代码(Country Code);
- 3 个字符码记录货币代码(Currency Code);
- 1 个字符码记录汇率指数(Currency Exponent);
- 4 个字符码记录周期授权金额(Amount Authorized per Cycle);
- 4 个字符码记录周期剩余金额(Amount Remaining this Cycle);
- 4 个字符码记录有效时间的起始周期(Cycle Begin Validity Date);
- 2 个字符码记录周期长度(Cycle Length);
- 1 个字符码记录重试次数(Retry Count);
- 6 个字符码记录可选项 PIN 码控制参数(PIN Control Parameters);
- 1 个字符码记录交换控制位(Interchange Controls);
- 2 个字符码记录 PAN 码服务限制位(PAN Service Restriction);
- 2 个字符码记录 SAN-1 码服务限制位(SAN-1 Service Restriction);
- 2 个字符码记录 SAN-2 码服务限制位(SAN-2 Service Restriction);
- 4 个字符码记录可选项截止日期(Expiration Date);
- 1 个字符码记录卡序列号(Card Sequence Number);
- 9 个字符码记录可选项卡片安全码(Card Security Number)。

Additional Data(额外数据):此域在银行卡系统中,可能会有第一、第二附属卡片账号数据,为可选项;1 个字符码用来记录中继标记;6 个字符码用来标记可选项密码校验数字。

Discretionary Data(授权数据):主要数据包含 1 个字符码用来记录 PIN 校验密钥索引码(PVKI);4 个字符码用来记录 PIN 校验值(PVV);3 个字符码用来记录卡校验值(CVV)或者卡校验码(CVC)。

如上所述磁道中数据编码格式可分成两种:磁道 1 是一种,由 6 个数据比特加 1 个奇偶校验位组成;磁道 2 和磁道 3 是一种,由 4 个数据比特加 1 个奇偶校验位组成。它们的字符集数据存储在磁性条纹中,格式并不是标准的 ASCII 字符集,而是由 ANSI 和 ISO 字符组成的集成字符集。

磁道 1 的最大容量为支持 79 个字母或者数字字符。磁道 1 的编码数据字符集包含 64 个数据,由 10 个阿拉伯数字(0~9)、26 个英文字母(A~Z)、3 个控制符(起始符、终止符、分隔符)和 25 个控制或特殊字符组成。磁道 1 的数据编码和格式如

4 MST磁安全传输支付系统

表4-4所列。

表4-4 磁道1的数据编码和格式

数据位						奇偶校验位	数值	信息	功能
b1	b2	b3	b4	b5	b6	b7			
0	0	0	0	0	0	1	0x00	空格	
1	0	0	0	0	0	0	0x01	!	
0	1	0	0	0	0	0	0x02	"	
1	1	0	0	0	0	1	0x03	#	特殊字符
0	0	1	0	0	0	0	0x04	$	
1	0	1	0	0	0	1	0x05	%	起始符
0	1	1	0	0	0	1	0x06	&	
1	1	1	0	0	0	0	0x07	'	
0	0	0	1	0	0	0	0x08	(	
1	0	0	1	0	0	1	0x09	)	
0	1	0	1	0	0	1	0x0A	*	特殊字符
1	1	0	1	0	0	0	0x0B	+	
0	0	1	1	0	0	1	0x0C	,	
1	0	1	1	0	0	0	0x0D	—	
0	1	1	1	0	0	0	0x0E	.	
1	0	0	1	0	0	1	0x0F	/	
0	0	0	0	1	0	0	0x10	0	
1	0	0	0	1	0	1	0x11	1	
0	1	0	0	1	0	1	0x12	2	
1	1	0	0	1	0	0	0x13	3	
0	0	1	0	1	0	1	0x14	4	
1	0	1	0	1	0	0	0x15	5	数字字符
0	1	1	0	1	0	0	0x16	6	
1	1	1	0	1	0	1	0x17	7	
0	0	0	1	1	0	1	0x18	8	
1	0	0	1	1	0	0	0x19	9	

续表 4-4

数据位						奇偶校验位	数值	信息	功能
b1	b2	b3	b4	b5	b6	b7			
0	1	0	1	1	0	0	0x1A	:	特殊字符
1	1	0	1	1	0	1	0x1B	;	
0	0	1	1	1	0	0	0x1C	<	
1	0	1	1	1	0	1	0x1D	=	
0	1	1	1	1	0	1	0x1E	>	
1	1	1	1	1	0	0	0x1F	?	终止符
0	0	0	0	0	1	0	0x20	@	特殊字符
1	0	0	0	0	1	1	0x21	A	英文字符
0	1	0	0	0	1	1	0x22	B	
1	1	0	0	0	1	0	0x23	C	
0	0	1	0	0	1	1	0x24	D	
1	0	1	0	0	1	0	0x25	E	
0	1	1	0	0	1	0	0x26	F	
1	1	1	0	0	1	1	0x27	G	
0	0	0	1	0	1	1	0x28	H	
1	0	0	1	0	1	0	0x29	I	
0	1	0	1	0	1	0	0x2A	J	
1	1	0	1	0	1	1	0x2B	K	
0	0	1	1	0	1	0	0x2C	L	
1	0	1	1	0	1	1	0x2D	M	
0	1	1	1	0	1	1	0x2E	N	
1	1	1	1	0	1	0	0x2F	O	
0	0	0	0	1	1	1	0x30	P	
1	0	0	0	1	1	0	0x31	Q	
0	1	0	0	1	1	0	0x32	R	
1	1	0	0	1	1	1	0x33	S	
0	0	1	0	1	1	0	0x34	T	
1	0	1	0	1	1	1	0x35	U	
0	1	1	0	1	1	1	0x36	V	
1	1	1	0	1	1	0	0x37	W	
0	0	0	1	1	1	0	0x38	X	
1	0	0	1	1	1	1	0x39	Y	
0	1	0	1	1	1	1	0x3A	Z	

续表 4-4

数据位						奇偶校验位	数值	信息	功能
b1	b2	b3	b4	b5	b6	b7			
1	1	0	1	1	1	0	0x3B	[	特殊字符
0	0	1	1	1	1	1	0x3C	\	特殊字符
1	0	1	1	1	1	0	0x3D	]	特殊字符
0	1	1	1	1	1	0	0x3E	^	分隔符
1	1	1	1	1	1	1	0x3F	_	特殊字符

磁道 1 的数据每字节由 7 比特构成,如果正向刷卡,则首先获得数据 b1,即该字节的最低位 bit0,然后依次为 b1、b2、b3、b4、b5、b6、b7,其中 b7 为该字节的奇偶校验位;如果逆向刷卡,则首先获得的是该字节的奇偶校验位。如图 4-10 所示为磁道 1 的正刷和反刷的顺序图。

磁道 1 正刷								
最高有效位							最低有效位	
b1	b2	b3	b4	b5	b6	b7	0	
磁道 1 反刷								
最高有效位							最低有效位	
0	b7	b6	b5	b4	b3	b2	b1	

图 4-10 磁道 1 的正刷和反刷的顺序图

磁道 2 的最大容量为支持 40 个数字字符,磁道 3 的最大容量为支持 107 个数字字符。磁道 2 编码数据字符集包含 16 个数据,由 10 个阿拉伯数字(0~9)、3 个控制符(起始符、终止符、分隔符)和 3 个控制或特殊字符组成。磁道 2 的数据编码和格式如表 4-5 所列。

磁道 2 和磁道 3 的数据每字节均由 5 比特构成,如果正向刷卡,则首先获得数据 b1,即该字节的最低位 bit0,然后依次为 b1、b2、b3、b4、b5,其中 b5 为该字节的奇偶校验位;如果逆向刷卡,则首先获得的是该字节的奇偶校验位。如图 4-11 所示为磁道 2 和磁道 3 的正刷和反刷的顺序图。

表 4-5 磁道 2 数据编码和格式

数据位				奇偶校验位	数 值	信 息	功 能
b1	b2	b3	b4	b5			
0	0	0	0	1	0x00	0	数字字符
1	0	0	0	0	0x01	1	
0	1	0	0	0	0x02	2	
1	1	0	0	1	0x03	3	
0	0	1	0	0	0x04	4	
1	0	1	0	1	0x05	5	
0	1	1	0	1	0x06	6	
1	1	1	0	0	0x07	7	
0	0	0	1	0	0x08	8	
1	0	0	1	1	0x09	9	
0	1	0	1	1	0x0A	:	特殊字符
1	1	0	1	0	0x0B	;	起始符
0	0	1	1	1	0x0C	<	特殊字符
1	0	1	1	0	0x0D	=	分隔符
0	1	1	1	0	0x0E	>	特殊字符
1	1	1	1	1	0x0F	?	终止符

磁道 2 和磁道 3 正刷							
最高有效位							最低有效位
b1	b2	b3	b4	b5	0	0	0
磁道 2 和磁道 3 反刷							
最高有效位							最低有效位
0	0	0	b5	b4	b3	b2	b1

图 4-11 磁道 2 和磁道 3 正刷的顺序图

以上主要介绍了数据格式、解码和卡信息示例,下面介绍一下数据在磁条上的存储和读取方式。每个记录的编码技术被称为双频记录。这种方法允许串行同步数据的记录。数据存储在磁条上通常采用 F2F(F2F 或者 F/2F Frequency/double

Frequency)双频制记录,编码包含数据和时钟转换,在两个时钟之间产生的磁通翻转记为"1",无磁通翻转记为"0"。数据按照字符的同步序列记录,中间不插入间隙,数据存储格式如图 4-12 所示。

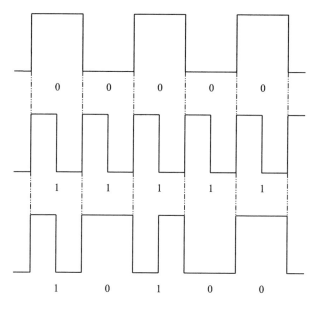

图 4-12　数据 F2F 存储格式

纵向冗余校验位

磁卡每个磁道的最后一个有效数据位为纵向冗余校验位,长度为一个字符码。该校验位对整帧的数据域进行校验,包含起始、结束、分割以及纵向冗余校验位本身,并不包含奇偶校验位,规则为每一字节对应的位上的 1 个数应为偶数个。它的取值应该是能使整轨数据进行"异或"后结果为零。所以对于磁道 1、2、3 上的信息其实都进行了双重校验,分别为字符码的奇校验和每轨整帧数据的 LRC 纵向冗余校验。

4.1.4　磁头原理

从 2015 年 6 月 30 日起,由第三方支付卡行数据安全标准(PCI DSS, Payment Card Industry Data Security Standard)发起的 PCI DSS 3.0 版本将由之前的一些可选要求的安全认证标识变成强制性的认证要求,PCI DSS 意在保护账户数据,为技术和操作要求提供一个基准。目前市面上主流应用的金融 POS 机,都陆续支持这个版本的标准,到目前为止最新版本为 2016 年 4 月发布的 PCI DSS 3.2。

第三方支付卡行数据安全标准 PCI DSS,是由 PCI 安全标准委员会的创始成员

Visa、Mastercard、American Express、Discover Financial Services、JCB 组织机构等制定的，主旨为推动国际上采用一致的数据安全措施，以全面保障交易安全。如表 4-6 所列为对所有涉及银行系统的信用卡信息机构做出的安全方面的标准，其中包括安全管理、策略、过程、网络体系结构、软件设计的要求等。

表 4-6 PCI 数据安全标准概述

建立并维护安全的网络和系统	安装并维护防火墙配置以保护持卡人数据
	不要使用供应商提供的默认系统密码和其他安全参数
保护持卡人数据	保护存储的持卡人数据
	加密持卡人数据在开放式公共网络中的传输
维护漏洞管理计划	为所有系统提供恶意软件防护并定期更新杀毒软件或程序
	开发并维护安全的系统和应用程序
实施强效访问控制措施	按业务知情需要限制对持卡人数据的访问
	识别并验证对系统组件的访问
	限制对持卡人数据的物理访问
定期监控并测试网络	跟踪并监控对网络资源和持卡人数据的所有访问
	定期测试安全系统和流程
维护信息安全政策	维护针对所有工作人员的信息安全政策

由五个不同组织发起的安全项目：Visa 持卡人信息安全项目、万事达站点数据保护项目、美国运通数据安全操作策略、Discover 的信息安全与规范、JCB 的数据安全项目。这几个组织都有一个大致相似的意图，就是创建一个发卡机构，提供额外水平的安全保护，确保商家满足最低限度的磁条或者芯片卡数据安全存储时，持卡人能处理和传送数据。在这种背景下，支付卡行业安全标准委员会（PCI SSC，Payment Card Industry Secutity Standards Council）成立，以组织创建统一的 PCI DSS 规范和标准。下面列出几个 PCI DSS 版本发布以及实施的重要时间节点：

➢ 2004 年 12 月 15 日，PCI DSS 1.0 版本发布；

➢ 2006 年 9 月，提供信息澄清和小版本 PCI DSS 1.1 更新；

➢ 2008 年 10 月 1 日，发布 PCI DSS 1.2 版本，该版本可增强透明度、提高灵活性、提升抗击不断变化的风险和威胁的能力；

➢ 2009 年 8 月，更新 PCI DSS 1.2 小版本，旨在创建更多的清晰性和一致性的标准以及支持文档说明等；

- 2010年10月,发布PCI DSS 2.0版本;
- 2013年11月,发布PCI DSS 3.0版本,激活和实施的时间段为从2014年1月1日到2015年6月30日;
- 2015年4月,发布PCI DSS 3.1版本,不过该版本在2016年10月31日之后就已经停止推荐实施了;
- 2016年4月,发布PCI DSS 3.2版本。

另外PCI DSS标准适用于所有涉及参与支付卡处理的所有实体,包括商户、处理商、收单机构、发卡机构和服务提供商。PCI DSS还适用于存储、处理或传输持卡人数据(CHD,Card Holder Data)或者敏感验证数据(SAD,Sensitive Authentication Data)的所有其他实体。PCI DSS包括一组保护持卡人信息的基本要求,并可能增加额外的管控措施以进一步管理和降低风险。如表4-7所列为PCI DSS 3.2账户数据中的持卡人数据和敏感验证数据,以及它们相对应的存储性和可读性的定义。

表4-7 金融POS机读头的安全要求

账户数据	数据元素	是否允许存储	存储数据是否可读
持卡人数据（CHD）	主账户（PAN）	是	是
	持卡人姓名		否
	业务码		
	失效日		
敏感验证数据（SAD）	全磁道数据	否	
	CAV2/CVC2/CVV2/CID		
	PIN/PIN 数据块		

根据PCI DSS标准,在金融POS机上需要保护的区域有:显示屏、安全键盘、刷卡核心区域、ISO/IEC 7816 IC接触卡片卡座、ISO/IEC 14443非接触卡片接口区域、磁头、主板与前壳、保护板与后壳等。本节重点介绍磁头原理,对其他组件的安全规则和策略不做过多介绍。对于POS机的磁头,第一是一般都需要采用加密磁头,这种磁头的PCBA上带有加密芯片,磁头与主板的连接器物理通道本身可以不做特别保护,但加密芯片会与主控制器之间有协商好的公私密钥对,分别存放在加密芯片以及主控器的安全代码库里,这样在进行磁头驱动数据读取时,双方都通过公私密钥对来生成会话密钥加密每次的通信数据;第二是基本上会采用三层板的设

计来封死磁头背面的焊点,其结构形式类似于核心保护区域,磁头连接器位于核心保护区域内。这样分别通过物理电路板的设计和数据通信加密的方式来提升整个主机端与磁头之间的信息安全等级,提高通过外部物理飞线方式去嗅探和窥测数据及控制总线的安全等级,从而达到一个基本安全和可控的系统设计。

由磁头的内部结构可知,磁头实际上是在一个本身并无磁性的铁芯上绕制线圈而成的,这就构成了一个电磁铁,具有磁通过切割线圈而产生电流和电流通过绕制在铁芯上的线圈而产生磁的两大特点。具体工作原理:电生磁是指在线圈上通过电流时,原来无磁性的铁芯就形成了有磁性的磁场,通过的电流消失后,磁场随即消失;磁生电则是指当有磁场靠近铁芯时,铁芯被磁化,铁芯就由无磁性变为有磁性,线圈中就产生感应电流,磁场消失后,线圈的感应电流随之消失。这就是制造磁头和磁头工作的最基本原理。如图 4-13 所示为纵向和垂直磁头的原理示意图。

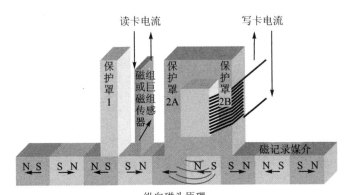

纵向磁头原理

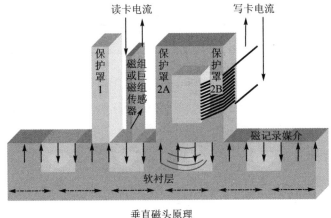

垂直磁头原理

图 4-13 磁头原理示意

1. 磁头数据读取原理

磁头数据读取,是利用电磁铁的磁耦合原理。当磁条卡吸收到磁力线后,磁力线通过线圈而产生感应电压。具体过程为:已做个人化数据的磁条卡上面的磁道已经按具体数据的变化规律被磁化了,磁道上便形成了间断的北磁极 N(North)、南磁极 S(South),这就是磁信息。刷卡时,磁卡上的磁道顺着工作缝隙运动,磁道上 N、S 磁极区就经过工作缝隙,反过来对磁头铁芯进行磁化,使没有磁性的铁芯产生感应磁场,形成穿过线圈的磁力线,线圈两端便产生感应电压,即电信号。电信号再通过磁头送到解码器运算电路中,把磁道中的字节信息一一解出,通过校验字符的奇校验以及磁道全帧的 LRC 纵向冗余校验,数据准确无误后上传到上层应用程序进行下一步信号处理。

2. 磁头数据写入原理

磁头数据写入,是利用电磁铁产生的电能去切割缠绕在其上的铁芯,从而产生磁场,磁场随着电能的变化,再将磁卡中的磁道数据进行修改或者写入。具体过程为:上层应用程序把需要写入的信息,进行数据协议包装或者重组,再通过磁数据译码器电路,把数据转换成磁卡上的可以识别的磁道数据,磁道数据下发到磁头,磁头随着数据变化形成磁卡上磁道可以识别的磁场信号,然后通过磁化作用把磁卡磁道中的数据进行改写或者灌入,于是上层下发的数据就变为磁信息储存到磁卡中了。

4.1.5 磁道解码应用

上面介绍了磁头如何把磁卡中对应的磁道数据在固定的时间内进行的磁信号变化,返回给上层的解码电路,解码电路根据磁卡数据对应磁道的数据格式,将磁信号转化为 0 和 1 的数据比特流输出的过程。

目前市场上硬件解码芯片供应商比较多,有南京博芯电子科技、美国 IDTECH 公司和台湾中青科技等。其中南京博芯电子科技提供的解码芯片主要有支持三轨磁道的解码芯片 PA110 和支持双轨磁道的解码芯片 PA1200,当然通过硬件连接方式它们还可以向下磁道兼容,例如 PA110 可以支持双轨磁道和单轨磁道,PA1200 可以向下兼容单轨磁道。美国 IDTECH 公司提供的磁卡解码芯片有单轨磁道解码芯片 ICC268R(21008 - 902RVA)系列、双轨磁道解码芯片 ICC198R(21008 - 900RVE)系列、三轨磁道解码芯片 ICC178R(1050LVVD)系列。台湾中青科技提供的芯片有单轨磁道解码芯片 M3-2100 系列、双轨磁道解码芯片 M3-2200 系列、三轨磁道解码芯片 M3-2300 和 M7-2300 系列。以上公司的产品信息仅为本书方便解释和对比产品用途，具体准确的产品信息需要与解码芯片的供应商去联系获取。

为方便说明磁道解码应用过程，本书选用南京博芯电子科技的 PA1100，封装 SOP8 作为示例进行说明，该芯片的特性如下：

- ➢ 支持单、双、三轨磁道的磁条卡数据解码；
- ➢ 内置 RAM 缓存数据功能；
- ➢ 硬件电路简单，无需额外的电阻电容耦合器件；
- ➢ 支持正反向磁条刷卡；
- ➢ 支持自动增益；
- ➢ 超低功耗，读卡电流小于 800 μA，休眠电流小于 1 μA；
- ➢ 读卡速度可达 10~254 cm/s；
- ➢ 读卡强度 30%~200%，符合 ISO 7811 标准；
- ➢ 采用 shift-out 串行接口；
- ➢ 芯片支持的供电范围为 2.7~3.6 V；
- ➢ 芯片封装支持 SOP8。

上面介绍了 PA1100 芯片无需额外的外围匹配电路，所以对于芯片的输入部分，只需与磁头的通道直接连接即可，此外还需必要的电源和地连接形成工作回路；对于芯片的输出部分，则通过 shift-out 串行接口与主机端接口直连即可，如图 4-14 所示为芯片内部框图以及硬件参考设计。

PA1100 SOP8 封装形式的引脚信号定义和说明如表 4-8 所列。

表 4-8 PA1100 SOP8 封装定义

序号	标识	说明	序号	标识	说明
1	VCOM	轨道公用基准	5	STROBE	控制信号
2	A	轨道 1	6	DATA	数据信号
3	B	轨道 2	7	VSS	地
4	C	轨道 3	8	VDD	电源

4 MST 磁安全传输支付系统

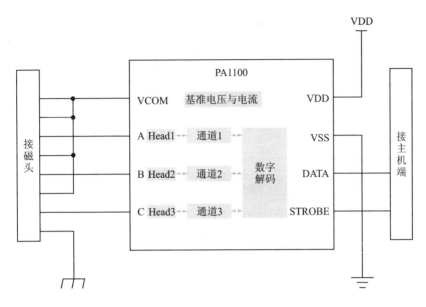

图 4-14 PA1100 内部框图

1. 磁卡 PA1100 解码驱动代码设计原理

（1）初始化设计

PA1100 在上电后需要初始化,初始化过程是由外接的 MCU(或其他主机端系统)驱动信号同时送给 PA1100 的 STROBE 和 DATA 引脚,它们两组信号的时序波形如图 4-15 所示。

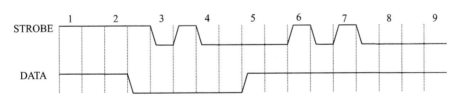

图 4-15 PA1100 初始化时序

PA1100 初始化时序的步骤如下：

① 对于 MCU(或其他主机端系统),连接到信号线 STROBE 和 DATA 的引脚应该配置成输出属性。

② 控制信号线 STROBE 和 DATA 同步保留至少约 900 ns(3×300 ns)的高电平状态。

③ DATA 信号拉成低电平后,再把 STROBE 信号变成低高低(010)电平变化,

81

其中的每一个低高低的逻辑电平至少持续 300 ns,然后 DATA 信号拉回到高电平的状态,此刻 STROBE 信号再变成高低高低(1010)电平变化。

④ 此刻就标志着 PA1100 芯片完成了初始化的过程,这时 DATA 信号应处于高电平状态,STROBE 信号应处于低电平状态。

具体使用可参考下面提供的 PA1100 解码驱动示例代码中的 void Pa1100_Reset(void) 函数。

(2) 读卡设计

初始化结束后,PA1100 进入等待刷卡状态。一旦有刷卡行为,PA1100 会把 DATA 信号拉低表示当前刷卡有效,当 MCU(或其他主机端系统)通过查询或者外部中断方式发现 DATA 信号被拉低后,驱动脉冲信号送给 STROBE 引脚,将存储在 PA1100 内的磁卡信息读取出来,时序波形如图 4-16 所示。STROBE 首先发一个高电平脉冲确认数据有效,然后循环发脉冲读取数据。PA1100 在 STROBE 下降沿将数据送出,MCU(或其他主机端系统)在 STROBE 为上升沿时将 DATA 信号线上的数据取走。两组信号的时序波形如图 4-16 所示。

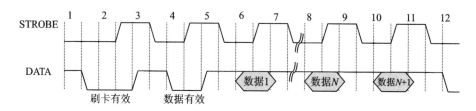

图 4-16　PA1100 读卡时序

PA1100 内部存储器,三个磁道总共有 704 bit×3＝2 112 bit 数据,数据存放顺序如下:

① 16 bit 前导码,软件处理时可忽略。

② 第一轨道数据,总长 704 bit,如果磁卡轨道信息不足 704 bit,后续数据补 1。

③ 第二轨道数据,总长 704 bit,如果磁卡轨道信息不足 704 bit,后续数据补 1。

④ 第三轨道数据,总长 704 bit,如果磁卡轨道信息不足 704 bit,后续数据补 1。

PA1100 读卡时序的步骤如下:

① 对于 MCU(或其他主机端系统),连接到信号线的 STROBE 引脚应该配置成输出属性,DATA 的引脚应该配置成输入属性(如果需要使用中断服务处理,那么还要设置好低电平触发中断的属性)。

② 当主机端系统通过查询或者中断方式侦测到 DATA 信号有拉低行为时,做一些延时和去抖处理,确认信号准确且有效后,即可表示此次刷卡有效,且主机端系统把 STROBE 信号从低拉回成高,可进入下一步骤。

③ 主机端系统通过控制 STROBE 信号发出低高逻辑电平(01)的变化后,去 DATA 信号线上读取数据,如果 DATA 信号线逻辑电平为低,那么表示 PA1100 解码芯片此次捕获的磁道数据有效,可以进入下一步骤。

④ 循环 16 次,每次循环都是主机端通过控制 STROBE 信号发出低高逻辑电平(01)的变化后,对 DATA 信号的逻辑电平进行读取,这些数据提供给前导码使用,在实际软件中可以直接忽略,然后进入正式的数据读取的步骤。

⑤ 分别对三个磁道循环 704 次,每次循环都是主机端通过控制 STROBE 信号发出低高逻辑电平(01)的变化后,把 DATA 信号对应的寄存器数据进行读取和处理,存入全局变量或者分配好的内存中,接下来就可以进入到最后的磁道用户数据分析,最终得到磁道中存储的用户应用数据。

具体使用可参考下面提供的 PA1100 解码驱动示例代码中的 Read_Track_Data() 和 Print_Track_Data() 函数。

(3) 结束设计

当主机端循环读取数据结束后,即在读完最后一个数据后,MCU(或其他主机端系统)再次把 STROBE 和 DATA 信号线配置成输出属性,并且驱动信号同时送给 STROBE 和 DATA,相当于重新初始化 PA1100,等 PA1100 初始化完成好后就可以等待下一次刷卡信息,如图 4-17 所示。

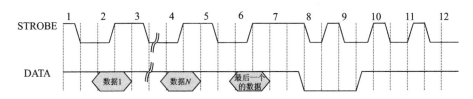

图 4-17 PA1100 结束和重新开始时序

PA1100 结束读卡和重新开始读卡设置的步骤如下:

① 当三个磁道的数据全部读取完毕时,在最后一帧数据结束后,MCU(或其他主机端系统)再次把 STROBE 和 DATA 信号线配置成输出属性,接管并且控制两组信号的逻辑电平变化。

② 主机端将控制信号线 STROBE 和 DATA 同步保留至少 900 ns(3×300 ns)的高电平状态。

③ DATA 信号拉成低电平后,再把 STROBE 信号变成低高低(010)电平变化,其中的每一个低高低的逻辑电平持续不低于 300 ns,然后 DATA 信号拉回到高电平状态,此刻 STROBE 信号再变成高低高低(1010)电平变化。

④ 此刻就标志着 PA1100 芯片完成了结束读卡和重新开始读卡设置的过程,这时 DATA 信号线应处于高电平状态,STROBE 信号应处于低电平状态。

具体使用可参考下面提供的 PA1100 解码驱动示例代码中的 void Pa1100_Reset(void)函数。

(4) 时序要求

以上所有时序说明中的 STROBE 信号的高、低电平时间持续不低于 300 ns。

2. 磁卡 PA1100 解码驱动示例代码

PA1100 芯片驱动代码的头文件 pa1100.h 如下:

```
/************************************************************
 *    Copyright (c) 2005 by National ASIC System Engineering Research Center.
 *    PROPRIETARY RIGHTS of ASIC are involved in the subject matter of this
 *    material. All manufacturing, reproduction, use, and sales rights
 *    pertaining to this subject matter are governed by the license agreement.
 *    The recipient of this software implicitly accepts the terms of the license.
 *
 *    File Name: PA1100.h
 *
 *    File Description:
 *             The file define some macro definition used in lcdc.c file.
 *
 *    Created by spone < spone_awp@yahoo.com.cn > , 2005 - 03 - 22
 ************************************************************/
#ifndef __PA1100_H
#define __PA1100_H

#include "sep4020.h"
#include "intc.h"
#include "sep4020_hal.h"
```

```c
U16 Get_Data;
static U8 error;
static U8 Direction;

#define REG32(addr)     (*(volatile unsigned int *)(addr))

#define delay(x)    {int i = x; while(i--);}

/*---------------------------------------------------
            SEP4020 GPIO port or any bit operation
---------------------------------------------------*/
#define SET_BIT(port, bit) REG32(port) |= (1 << bit)
#define CLR_BIT(port, bit) REG32(port) &= (~(1 << bit))


/*---------------------------------------------------*/
/******************************************************/
/* SEP4020_PA1100 */
/******************************************************/
#define PA1100_DATA_BIT         10
#define PA1100_STROBE_BIT       0

#define PA1100_DATA_SEL         GPIO_PORTG_SEL
#define PA1100_DATA_DIR         GPIO_PORTG_DIR
#define PA1100_DATA_DATA        GPIO_PORTG_DATA

#define PA1100_STROBE_SEL       GPIO_PORTA_SEL
#define PA1100_STROBE_DIR       GPIO_PORTA_DIR
#define PA1100_STROBE_DATA      GPIO_PORTA_DATA

#define PA1100_DATA_BIT_H    (*(RP)PA1100_DATA_DATA |= 0x1 << PA1100_DATA_BIT)
#define PA1100_DATA_BIT_L    (*(RP)PA1100_DATA_DATA &= (~0x1 << PA1100_DATA_BIT))

#define PA1100_STROBE_BIT_H  (*(RP)PA1100_STROBE_DATA |= 0x1 << PA1100_STROBE_BIT)
#define PA1100_STROBE_BIT_L  (*(RP)PA1100_STROBE_DATA &= (~0x1 << PA1100_STROBE_BIT))

#define PA1100_DATA    ((*(RP)PA1100_DATA_DATA & 0x400) >> PA1100_DATA_BIT)
```

```c
#define PA1100_DATA_IN     (*(RP)PA1100_DATA_DIR |= 0x1 << PA1100_DATA_BIT)

#define PA1100_DATA_OUT    (*(RP)PA1100_DATA_DIR &= (~0x1 << PA1100_DATA_BIT))

void Pa1100_Port_Init(void);
U8 Track_Data_Decode(U8 data);
void Pa1100_Reset(void);
void Track1_Check(void);
void Track2_Check(void);
void Track3_Check(void);
void Print_Track_Data(void);
void Track_Data_Reverse_Handle(void);
void Read_Track_Data(void);

#endif
```

驱动代码示例 pa1100.c 如下：

```c
/*********************************************************
 *   Copyright (c) 2009 by PROCHIP Limited.
 *   PROPRIETARY RIGHTS of PROCHIP Limited are involved in the subject matter of this
 *   material. All manufacturing, reproduction, use, and sales rights
 *   pertaining to this subject matter are governed by the license agreement.
 *   The recipient of this software implicitly accepts the terms of the license.
 *
 *   File Name: pa1100.c
 *
 *   File Description: pa1100 的函数文件
 *
 *   Version        Date           Author
 *   ------------------------------------------------------
 *   1.0            2012.11.15     lori
 *
 *********************************************************/
#include <stdio.h>
#include <string.h>
#include "pa1100.h"

#define time    300
```

4 MST 磁安全传输支付系统

```c
U8 CardData[380];

U8 AsciiData[] = {
0xff,0xff,' ','!',''','#','$','%','&','? ','(',')','*','+',',','-','.','/',
'0','1','2','3','4','5','6','7','8','9',':',';',' < ',' = ',' > ','? ',
'@','A','B','C','D','E','F','G','H','I','J','K','L','M','N','O',
'P','Q','R','S','T','U','V','W','X','Y','Z','[','\\',']','^','_',
'0','1','2','3','4','5','6','7','8','9',':',';',' < ',' = ',' > ','? '};

U8 MagData[] = {
0x00,0x80,0x40,0x01,0x02,0x43,0x04,0x45,0x46,0x07,0x08,0x49,0x4A,0x0B,0x4C,0x0D,
0x0E,0x4F,
0x10,0x51,0x52,0x13,0x54,0x15,0x16,0x57,0x58,0x19,0x1A,0x5B,0x1C,0x5D,0x5E,0x1F,
0x20,0x61,0x62,0x23,0x64,0x25,0x26,0x67,0x68,0x29,0x2A,0x6B,0x2C,0x6D,0x6E,0x2F,
0x70,0x31,0x32,0x73,0x34,0x75,0x76,0x37,0x38,0x79,0x7A,0x3B,0x7C,0x3D,0x3E,0x7F,
0x90,0x81,0x82,0x93,0x84,0x95,0x96,0x87,0x88,0x99,0x9A,0x8B,0x9C,0x8D,0x8E,
0x9F};

/**************************************************
函数原型:U8 Track_Data_Decode(U8 data)
入口参数:data:要译码的数据
返回参数:data:译码后的数据
函数功能:轨道数据译码
**************************************************/
U8 Track_Data_Decode(U8 data)
{
    U16 temp = 0;

    data = ~data;
    while(temp < 82)
    {
        //check table to find ASCII
        if(data == MagData[temp])
        {
            data = AsciiData[temp];
            break;
        }
        temp++;
```

```
        }

        if(temp == 82)
        {
            data = 0xff;
            if(AsciiData ! = 0x00)
                error = temp;
        }
        return data;
}


/*************************************************************
函数原型:void Pa1100_Port_Init(void)
入口参数:无
返回参数:无
函数功能:配置 IO 端口
*************************************************************/
void Pa1100_Port_Init(void)
{
        SET_BIT(PA1100_DATA_SEL, PA1100_DATA_BIT);
        CLR_BIT(PA1100_DATA_DIR, PA1100_DATA_BIT);

        SET_BIT(PA1100_STROBE_SEL, PA1100_STROBE_BIT);
        CLR_BIT(PA1100_STROBE_DIR, PA1100_STROBE_BIT);
}


/*************************************************************
函数原型:void Pa1100_Reset(void)
入口参数:无
返回参数:无
函数功能:PA1100 复位
*************************************************************/
void Pa1100_Reset(void)
{
        PA1100_DATA_BIT_H;
        PA1100_STROBE_BIT_H;
        delay(300);
```

```
        PA1100_DATA_BIT_L;
        delay(100);
        PA1100_STROBE_BIT_L;
        delay(100);

        PA1100_STROBE_BIT_H;
        delay(100);
        PA1100_STROBE_BIT_L;
        delay(300);
        PA1100_DATA_BIT_H ;
        delay(100);

        PA1100_STROBE_BIT_H;
        delay(100);
        PA1100_STROBE_BIT_L;
        delay(100);
        PA1100_STROBE_BIT_H;
        delay(100);
        PA1100_STROBE_BIT_L;
        delay(100);
}


/***********************************************************
函数原型:void Track1_Check(void)
入口参数:无
返回参数:无
函数功能:轨道1数据处理
***********************************************************/
void Track1_Check(void)
{
    U8 tmp = 0, temp = 0, k = 0;
    U16 i = 0, x = 0;

    if(((CardData[0] == 0xBA) && (CardData[1] != 0x83))
        Direction = 1;
    else
        Direction = 0;
```

```c
//LRC Check
if(Direction)
{
    for(i = 0; i < 100; i++)
    if((CardData[i] == 0xFF) && (CardData[i + 1] == 0xFF))
    break;
    x = i;
    for(i = 0; i < x - 1; i++)
    tmp = tmp ^ (~CardData[i]);
    if((tmp & 0x3f) != ((~CardData[x - 1]) & 0x3f))
    error = 1;
}

if(!Direction)
{
    for(i = 0; i < 100; i++)
    if((CardData[i] == 0xFF) && (CardData[i + 1] == 0xFF))
    break;
    x = i;
    for(k = 0; k < 7; k++)
    if((CardData[i - 1] & (0x40 >> k)) == 0x00)
    break;

    for(; i > 0; i--)
    {
        tmp = (CardData[i] << k) & 0x7F;
        temp = (CardData[i - 1] & 0x7F) >> (7 - k);
        tmp |= temp;
        CardData[i] = tmp | 0x80;
    }

    if(k != 0)
    {
        tmp = (CardData[0] << k) & 0x7F;
        temp = 0;
        while(k--)
        {
            temp <<= 1;
```

```
                    temp |= 0x01;
                }
                tmp |= temp;
                CardData[0] = tmp | 0x80;
            }
        //LRC Check
        if(x != 0)
        {
            tmp = 0;
            for(i = 1; i < x; i++)
                tmp ^= (~CardData[i]);
            if((tmp & 0x7e) != ((~CardData[0]) & 0x7e))
                error = 1;
        }
    }
}


/***********************************************************
函数原型:void Track2_Check(void)
入口参数:无
返回参数:无
函数功能:轨道2数据处理
***********************************************************/
void Track2_Check(void)
{
    U8 tmp = 0, temp = 0, k = 0;
    U16 i = 0, x = 0;

    if(!Direction)
    if((CardData[100] == 0x74) && (CardData[101] != 0x60))
        Direction = 1;

    //LRC Check
    if(Direction)
    {
        for(i = 100; i < 240; i++)
        if((CardData[i] == 0x7F) && (CardData[i+1] == 0x7F))
            break;
```

```
            x = i;

        for(i = 100; i < x - 1; i++ )
            tmp ^= (~CardData[i]);
            if((tmp & 0x0F) != ((~CardData[x - 1]) & 0x0F))
            error = 1;
    }

    if(!Direction)
    {
        for(i = 100; i < 240; i++ )
            if((CardData[i] == 0x7F) && (CardData[i + 1] == 0x7F))
            break;
            x = i;

        for(k = 0; k < 5; k++ )
            if((CardData[i - 1] & (0x10 >> k)) == 0x00)
            break;

        for(; i > 100; i-- )
        {
            tmp = (CardData[i] << k) & 0x1F;
            temp = (CardData[i - 1] & 0x1F) >> (5 - k);
            tmp |= temp;
            CardData[i] = tmp | 0x60;
        }

        if(k != 0)
        {
            tmp = (CardData[100] << k) & 0x1F;
            temp = 0;
            while(k-- )
            {
                temp <<= 1;
                temp |= 0x01;
            }
            tmp |= temp;
            CardData[100] = tmp | 0x60;
```

```c
            }
            //LRC Check
            if(x != 100)
            {
                tmp = 0;
                for(i = 101; i < x; i++)
                    tmp ^= (~CardData[i]);
                if((tmp & 0x1e) != ((~CardData[100]) & 0x1e))
                    error = 1;
            }
        }
    }


/****************************************************
函数原型:void Track3_Check(void)
入口参数:无
返回参数:无
函数功能:轨道 3 数据处理
****************************************************/
void Track3_Check(void)
{
    U8 tmp = 0, temp = 0, k = 0;
    U16 i = 0, x = 0;

    if(!Direction)
    if(((CardData[100] == 0x74) && (CardData[101] != 0x60)))
    Direction = 1;

    //LRC Check

    if(Direction)
    {
        for(i = 240; i < 380; i++)
        if((CardData[i] == 0x7F) && (CardData[i + 1] == 0x7F))
        break;
        x = i;
        for(i = 240; i < x - 1; i++)
        tmp = tmp^(~CardData[i]);
```

```
            if((tmp & 0x0F) ! = ((~CardData[x - 1]) & 0x0F))
            error = 1;
    }

    if(!Direction)
    {
        for(i = 240; i < 380; i ++ )
        if((CardData[i] == 0x7F) && (CardData[i + 1] == 0x7F))
        break;
        x = i;

        for(k = 0; k < 5; k ++ )
        if((CardData[i - 1] & (0x10 >> k)) == 0x00)
        break;

        for(; i > 240; i -- )
        {
            tmp = (CardData[i] << k) & 0x1F;
            temp = (CardData[i - 1] & 0x1F) >> (5 - k);
            tmp |= temp;
            CardData[i] = tmp | 0x60;
        }

        if(k ! = 0)
        {
            tmp = (CardData[240] << k) & 0x1F;
            temp = 0;
            while(k -- )
            {
                temp << = 1;
                temp |= 0x01;
            }
            tmp |= temp;
            CardData[240] = tmp | 0x60;
        }

    }
    //LRC Check
    if(x ! = 240)
```

```c
        {
            tmp = 0;
            for(i = 241; i < x; i++)
                tmp ^= (~CardData[i]);
            if((tmp & 0x1e) != ((~CardData[240]) & 0x1e))
                error = 1;
        }
    }
}


/***************************************************************
函数原型:void Print_Track_Data(void)
入口参数:无
返回参数:无
函数功能:输出轨道数据
***************************************************************/
void Print_Track_Data(void)
{
    U16 i = 0;

    if(Direction)
    {
        printf("Positive charge:\n");

        printf("The track1 data is:\n");
        for(i = 0; i < 100; i++)
        {
            CardData[i] = Track_Data_Decode(CardData[i]);
            if(CardData[i] != 0xff)
                printf("%c", CardData[i]);
        }
        printf("\n");

        printf("The track2 data is:\n");
        for(i = 100; i < 240; i++)
        {
            CardData[i] = Track_Data_Decode(CardData[i]);
            if(CardData[i] != 0xff)
```

```c
            printf(" %c", CardData[i]);
        }
        printf("\n");

        printf("The track3 data is:\n");
        for(i = 240; i < 380; i++)
        {
            CardData[i] = Track_Data_Decode(CardData[i]);
            if(CardData[i] != 0xff)
                printf(" %c", CardData[i]);
        }
    }
    else
    {
        printf("Reverse charge:\n");

        Track_Data_Reverse_Handle();

        printf("The track1 data is:\n");
        for(i = 1; i < 101; i++)
        {
            CardData[100 - i] = Track_Data_Decode(CardData[100 - i]);
            if(CardData[100 - i] != 0xff)
                printf(" %c", CardData[100 - i]);
        }

        printf("\n");

        printf("The track2 data is:\n");
        for(i = 141; i < 281; i++)
        {
            CardData[380 - i] = Track_Data_Decode(CardData[380 - i]);
            if(CardData[380 - i] != 0xff)
                printf(" %c", CardData[380 - i]);
        }

        printf("\n");
```

```c
        printf("The track3 data is:\n");
        for(i = 1; i < 141; i++)
        {
            CardData[380 - i] = Track_Data_Decode(CardData[380 - i]);
            if(CardData[380 - i] != 0xff)
                printf(" %c", CardData[380 - i]);
        }
    }
    printf("\n");
}


/***********************************************************
函数原型:void Track_Data_Reverse_Handle(void)
入口参数:无
返回参数:无
函数功能:反向刷卡数据处理
***********************************************************/
void Track_Data_Reverse_Handle(void)
{
    U16 i = 0;
    U8 tmp = 0;

    for(i = 0; i < 100; i++)
    {
        tmp = (CardData[i] >> 6) & 0x01;
        tmp |= (CardData[i] >> 4) & 0x02;
        tmp |= (CardData[i] >> 2) & 0x04;
        tmp |= CardData[i] & 0x08;
        tmp |= (CardData[i] << 2) & 0x10;
        tmp |= (CardData[i] << 4) & 0x20;
        tmp |= (CardData[i] << 6) & 0x40;
        CardData[i] = tmp | 0x80;
    }

    for(; i < 240; i++)
    {
        tmp = (CardData[i] >> 4) & 0x01;
        tmp |= (CardData[i] >> 2) & 0x02;
```

```c
            tmp |= CardData[i] & 0x04;
            tmp |= (CardData[i] << 2 ) & 0x08;
            tmp |= (CardData[i] << 4) & 0x10;
            CardData[i] = tmp | 0x60;
        }

        for(; i < 380; i++)
        {
            tmp = (CardData[i] >> 4) & 0x01;
            tmp |= (CardData[i] >> 2) & 0x02;
            tmp |= CardData[i] & 0x04;
            tmp |= (CardData[i] << 2) & 0x08;
            tmp |= (CardData[i] << 4) & 0x10;
            CardData[i] = tmp | 0x60;
        }
    }


/************************************************************
函数原型:void Read_Track_Data(void)
入口参数:无
返回参数:无
函数功能:读取刷卡数据
************************************************************/
void Read_Track_Data(void)
{
    int i = 0, tmp = 0, temp = 0;

    Get_Data = 0;
    error = 0;

    printf("track1:\n");
    //read strip 1
    for(i = 0; i < 704; i++)
    {
        PA1100_STROBE_BIT_L;
        delay(time);

        temp = (PA1100_DATA << 6) & 0x40;
```

```c
        tmp = (tmp >> 1) & 0x3f;
        tmp |= temp;
        if (((i + 1) % 7) == 0)
        {
            CardData[Get_Data] = tmp | 0x80;
            printf(" % x ", CardData[Get_Data]);
            Get_Data ++ ;
            tmp = 0;
        }

        PA1100_STROBE_BIT_H;
        delay(time);
}
printf("\n");
printf("Get_Data = % d\n", Get_Data);

Track1_Check();

printf("track2:\n");
//read strip 2
for (i = 0; i < 704; i ++ )
{
    PA1100_STROBE_BIT_L;
    delay(time);

    temp = (PA1100_DATA << 4) & 0x10;
    tmp = (tmp >> 1) & 0x0f;
    tmp |= temp;
    if (((i + 1) % 5) == 0)
    {
        CardData[Get_Data] = tmp | 0x60;
        printf(" % x ", CardData[Get_Data]);
        Get_Data ++ ;
        tmp = 0;
    }

    PA1100_STROBE_BIT_H;
    delay(time);
```

```
    }
    printf("\n");
    printf("Get_Data = %d\n", Get_Data);

    Track2_Check();

    printf("track3:\n");
    //read strip 3
    for(i = 0;i < 704;i++)
    {
        PA1100_STROBE_BIT_L;
        delay(time);

        temp = (PA1100_DATA << 4) & 0x10;
        tmp = (tmp >> 1) & 0x0f;
        tmp |= temp;
        if(((i+1) % 5) == 0)
        {
            CardData[Get_Data] = tmp | 0x60;
            printf(" %x ", CardData[Get_Data]);
            Get_Data++;
            tmp = 0;
        }

        PA1100_STROBE_BIT_H;
        delay(time);
    }
    printf("\n");
    printf("Get_Data = %d\n", Get_Data);

    Track3_Check();
}

void Pa1100_Test(void)
{
    int i;

    for(i = 0; i < 380; i++)
```

```c
        CardData[i] = 0;

    Pa1100_Port_Init();         //初始化 IO 端口

    Pa1100_Reset();             //初始化 PA1100

    PA1100_DATA_IN;

//while(1)
{
    delay(1000);

    //wait card swipe
    printf("\nWaiting for credit card...\n");
    while(PA1100_DATA == 1)
    {
        delay(time);
    }
    printf("\nCharge effective!\n");

    delay(5000);

    PA1100_STROBE_BIT_H;
    delay(time);

    //check data high -- CP finish
    while(PA1100_DATA == 0)
    {
        delay(time);
    }

    //wait BR
    PA1100_STROBE_BIT_L;
    delay(1000);
    PA1100_STROBE_BIT_H;

    //check data high -- BR finish
    while(PA1100_DATA == 0)
```

```
    {
        delay(time);
    }

    delay(time);

    //read 16bit preamble
    for (i = 0; i < 16; i++)
    {
        PA1100_STROBE_BIT_L;
        delay(time);
        PA1100_STROBE_BIT_H;
        delay(time);
    }

    Read_Track_Data();

    //if(!Direction)
        //Track_Data_Reverse_Handle();

    Print_Track_Data();
    }
}
```

4.2 MST 磁传输工作原理

前面已经详细介绍了传统磁条卡、磁道和磁头的工作原理，并提供了读卡应用程序作为参考。其实 MST 的电路原理和实现过程，本质上和磁头去读磁条卡的磁道原理是一样的，只是后者需要通过接触的方式，将卡片的磁力线再传到磁头上去，从而把数据读回来；而前者则相当于磁力线的能量通过驱动放大后，传递到磁头，然后磁头吸收回来后上传原始的解码电路，把数据正确解析出来。目前大部分的磁条卡业务应用，并没有把三个磁道的功能都使用上，主要还是在 1 和 2 个磁道之间进行工作，所以对于大部分的 MST 使用场景，原来商户的基础设施并不需要进行修改或者升级，但是有些支付场景下的基础设备还是需要更新 POS 端的系统才可以顺利支持 MST 方式支付。

4 MST 磁安全传输支付系统

电磁是在 1820 年由一位丹麦的物理学家汉斯·克里斯蒂安·奥斯特（Hans Christian Oersted,1777—1851 年）发现的。他在使用一些电子设备做实验时,发现当给一根电线通过强电流时,附近的指南针会发生偏转,不再指向北方,奥斯特意识到是电流正在产生磁力,扰乱了指南针指针的方向,于是他发现了电磁。理论上讲,任何电流总是能产生磁场的,以这种方式产生的磁叫电磁,产生的磁铁叫电磁铁,磁场以这种方式产生可以做到许多永久磁铁做不到的事情。因为可以通过开关电子电路来控制电磁场的开关状态,也可以通过控制电流的大小去控制电磁场的强弱,所以 MST 电路的设计也是基于最基本的电磁原理。

MST 电磁发送电路与磁头读卡电路相比,其本质上有些相似,只是前者需要把磁道的数据译码成差分曼彻斯特 F2F 编码,然后再通过驱动电路带动线圈产生电磁场,从而把数据发送出去;而后者比较像一个相反的过程,即磁头通过接收磁卡中的磁化感应到磁头,磁头再把信号送回到差分曼彻斯特 F2F 解码器,解码器把最终的数据解出来。前面已经具体介绍过 F2F 编码的原理,图 4-18 所示为磁头在读取外部磁卡或者接收 MST 信号后再与其对应的 F2F 编码示意图。

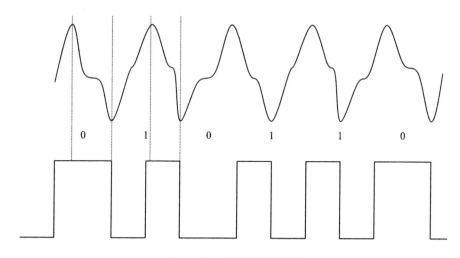

图 4-18 F2F 编码示意图

如前面磁头工作原理的介绍,磁卡中磁道的读/写都是由磁头来完成的,为了将数据信息写入到磁卡中,首先需要对信息进行编码,然后把编码后的信号转化成电流,通过磁头与磁卡的磁道面贴近,磁头与磁卡以一定的速度进行相对运动,磁道被磁化,信息就被写入到磁卡的磁道中了。在调频制记录方式中,信息的写入是依赖于写入电流频率的变化来实现的,而 F2F 制编码则是在 FM 调频制的基础上的一种改进的编码格式,其编码规则就是在记录数据"1"时,写电流在位周期中间改变方向;记录"0"时,写电流不改变方向。写电流改变方向时发生在每次位周期的边界。

磁卡中磁道数据的读出是磁道数据写入的一个反向过程,磁头在接触到磁卡中的磁道或者 MST 信号时,使磁头上的磁路发生了磁通变化,并根据电磁感应定律,磁头线圈产生感应电势,这个过程磁数据转换成了电信号,磁头线圈两端产生电压信号,解码电路通过解码数据后,完成了磁卡信息的读出过程。

目前,磁卡领域使用的基本都是差分曼彻斯特编码,可采用的编码方式还有:调频制(FM,Frequency Modulation)、调相制(PM,Phase Modulation)和改进调频制式(MFM,Modified Frequency Modulation)。其中改进调频制式 MFM 主要还是为了在不提高物理磁卡磁道密度的前提下能提高两倍的记录数据密度量。

MST 的应用与磁卡应用的主要区别就是前者可以通过类似非接触磁头的方式进行通信,后者则要通过磁头接触磁卡磁道的方式进行数据交换,但是其本质的电磁原理还是一样的。MST 的线圈靠近磁头时,与传统的 POS 机磁头线圈产生了互感,从而实现非接触方式的电磁通信。

为方便说明工作原理,这里我们以恩智浦公司的单片机 LPC5410 作为主机端例子来实现 F2F 编码数据格式,然后通过一颗双通道的 H 桥电机驱动芯片连接到线圈进行输出,这颗驱动芯片以德州仪器公司的 DRV8833 为例。如图 4-19 所示为 MST 硬件连接示意框图。

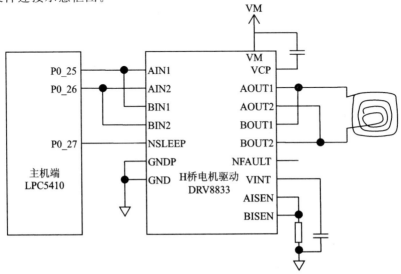

图 4-19 MST 硬件连接示意框图

4 MST 磁安全传输支付系统

主机端 LPC5410 WLCSP49 封装形式的 P0_25、P0_26 和 P0_27 引脚信号定义和说明如表 4-9 所列。

表 4-9 LPC5410 WLCSP49 封装定义

序号	标识	说明
E2	P0_25	端口 0 中的通用逻辑输入/输出 25 引脚
E1	P0_26	端口 0 中的通用逻辑输入/输出 26 引脚
D2	P0_27	端口 0 中的通用逻辑输入/输出 27 引脚

DRV8833 16 引脚的 HTSSOP 封装形式的引脚信号定义和说明如表 4-10 所列。

表 4-10 DRV8833 HTSSOP 封装定义

序号	标识	说明
1	NSLEEP	逻辑高为使能这个芯片进入正常的工作模式,逻辑低为让芯片进入低功耗模式,并且在此模式下芯片的内部逻辑状态将会复位一次
2	AOUT1	A 桥输出 1
3	AISEN	A 桥的地或者串联到电流检测电阻
4	AOUT2	A 桥输出 2
5	BOUT2	B 桥输出 2
6	BISEN	B 桥的地或者串联到电流检测电阻
7	BOUT1	B 桥输出 1
8	NFAULT	错误输出,当超过工作温度或者过流时,送出低电平给主机端(此文中提供的框架设计并没有连接这个信号到主机端,实际在做 MST 应用设计时推荐考虑加入连接到主机端)
9	BIN1	B 桥输入 1
10	BIN2	B 桥输入 2
11	VCP	门驱动电压,通过电容串联到 VM 信号上
12	VM	芯片工作电压输入
13	GND	芯片接地
13	GNDPPAD	芯片接地
14	VINT	芯片内部旁路设计,串联电容到地
15	AIN2	A 桥输入 2
16	AIN1	A 桥输入 1

DRV8833 中的 AIN1 和 AIN2 输入针对控制 AOUT1 和 AOUT2 输出的状态；而 BIN1 和 BIN2 输入则针对控制 BOUT1 和 BOUT2 输出的状态。如图 4-19 所示的 MST 硬件连接示意框图，AIN1&BIN1 和 AOUT1&BOUT1 并联在一起，AIN2&BIN2 和 AOUT2&BOUT2 并联在一起，所以实际上它们的逻辑控制关系只有四种，如表 4-11 所列为它们之间的逻辑控制关系。

表 4-11　DRV8833 驱动芯片设计的逻辑电平参考

AIN1&BIN1	AIN2&BIN2	AOUT1&BOUT1	AOUT2&BOUT2
0	0	高阻态	高阻态
0	1	低电平状态	高电平状态
1	0	高电平状态	低电平状态
1	1	低电平状态	低电平状态

主机端 LPC5410 通过控制通用的数字电平逻辑引脚 P0_25、P0_26 和 P0_27，从而控制 A 桥和 B 桥的能量输出，以达到控制连接的后端线圈电路去输出电磁信号的目的。上面描述过的 F2F 编码原理，写电流改变方向即意味着线圈上的电势改变方向，MST 的线圈由 DRV8833 控制，当在某一段时间内，AIN1 输出 1，同时 AIN2 输出 0 时，线圈中的 AOUT1&BOUT1 被连接到电源电平，而 AOUT2&BOUT2 则被连接到地，线圈中产生从 AOUT1&BOUT1 到 AOUT2&BOUT2 的电流，并产生对应方向的电磁场；当在另一段时间内，AIN1 输出 0，同时 AIN2 输出 1 时，线圈中的 AOUT1&BOUT1 被连接到地，而 AOUT2&BOUT2 被连接到电源电平，线圈中产生从 AOUT2&BOUT2 到 AOUT1&BOUT1 的电流，电势和电流改变了方向，于是也就产生了反向的电磁场。

根据 F2F 制编码的标准，当 MST 发送二进制数据"0"时，需要改变一次线圈中的电流方向并保持一段时间不变，该段时间就为一个完整的位周期；当 MST 发送二进制数据"1"时，需要改变一次线圈中电流方向并维持 1/2 位周期的时间不变，然后再改变一次线圈中的电流方向并维持 1/2 位周期的时间不变。

正常在标准的银行卡磁卡中，产生的 F2F 编码数据信号的位周期为 500 μs 左右，考虑到 MST 是通过非接触感应的方式传递磁信号给磁头的，过程中有可能会有传输时间消耗，所以我们这里就暂时把位周期时间设定在 400 μs 内，图 4-19 中主机端 LPC5410 的 P0_25 去控制 AIN1 和 BIN1，P0_26 去控制 AIN2 和 BIN2，那么想

4 MST 磁安全传输支付系统

要发送"101"的数据可以参考如下的代码设计:

```c
#define LPC_IOCON_BASE              0x4001C000UL
#define LPC_GPIO_PORT_BASE          0x1C000000UL

#define __STATIC_INLINE static inline

#ifdef __cplusplus
#define   __I  volatile              /*!< Defines 'read only' permissions */
#else
#define   __I  volatile const        /*!< Defines 'read only' permissions */
#endif
#define   __O  volatile              /*!< Defines 'write only' permissions */
#define   __IO volatile              /*!< Defines 'read/write' permissions */

typedef unsigned int uint32_t;

/**
 * @brief LPC5410X IO Configuration Unit register block structure
 */
typedef struct {
    __IO uint32_t PIO[2][32];
    /*!< LPC5410X IOCON Structure */
} LPC_IOCON_T;

/**
 * @brief Array of IOCON pin definitions passed to Chip_IOCON_SetPinMuxing() must be in this format
 */
typedef struct {
    uint32_t port : 8;            /* Pin port */
    uint32_t pin : 8;             /* Pin number */
    uint32_t modefunc : 16;       /* Function and mode */
} PINMUX_GRP_T;

/**
 * @brief GPIO port register block structure
 */
typedef struct {
```

```c
    /*!< GPIO_PORT Structure */
    __IO uint8_t B[128][32];
/*!< Offset 0x0000: Byte pin registers ports 0 to n; pins PIOn_0 to PIOn_31 */

    __IO uint32_t W[32][32];
/*!< Offset 0x1000: Word pin registers port 0 to n */

    __IO uint32_t DIR[32];
/*!< Offset 0x2000: Direction registers port n */

    __IO uint32_t MASK[32];
/*!< Offset 0x2080: Mask register port n */

    __IO uint32_t PIN[32];
/*!< Offset 0x2100: Portpin register port n */

    __IO uint32_t MPIN[32];
/*!< Offset 0x2180: Masked port register port n */

    __IO uint32_t SET[32];
/*!< Offset 0x2200: Write: Set register for port n Read: output bits for port n */

    __O uint32_t CLR[32];
/*!< Offset 0x2280: Clear port n */

    __O uint32_t NOT[32];
/*!< Offset 0x2300: Toggle port n */

} LPC_GPIO_T;

/**
 * IOCON function and mode selection definitions
 * See the User Manual for specific modes and functions supported by the
 * various LPC15XX pins.
 */
#define IOCON_FUNC0             0x0
/*!< Selects pin function 0 */
```

```
#define IOCON_FUNC1              0x1
/*!< Selects pin function 1 */

#define IOCON_FUNC2              0x2
/*!< Selects pin function 2 */

#define IOCON_FUNC3              0x3
/*!< Selects pin function 3 */

#define IOCON_FUNC4              0x4
/*!< Selects pin function 4 */

#define IOCON_FUNC5              0x5
/*!< Selects pin function 5 */

#define IOCON_FUNC6              0x6
/*!< Selects pin function 6 */

#define IOCON_FUNC7              0x7
/*!< Selects pin function 7 */

#define IOCON_MODE_INACT         (0x0 << 3)
/*!< No addition pin function */

#define IOCON_MODE_PULLDOWN      (0x1 << 3)
/*!< Selects pull-down function */

#define IOCON_MODE_PULLUP        (0x2 << 3)
/*!< Selects pull-up function */

#define IOCON_MODE_REPEATER      (0x3 << 3)
/*!< Selects pin repeater function */

#define IOCON_HYS_EN             (0x1 << 5)
/*!< Enables hysteresis */
#define IOCON_GPIO_MODE          (0x1 << 5)
/*!< GPIO Mode */
#define IOCON_I2C_SLEW           (0x1 << 5)
```

```c
/*!< I2C Slew Rate Control */
#define IOCON_INV_EN            (0x1 << 6)
/*!< Enables invert function on input */
#define IOCON_ANALOG_EN         (0x0 << 7)
/*!< Enables analog function by setting 0 to bit 7 */
#define IOCON_DIGITAL_EN        (0x1 << 7)
/*!< Enables digital function by setting 1 to bit 7(default) */
#define IOCON_STDI2C_EN         (0x1 << 8)
/*!< I2C standard mode/fast-mode */
#define IOCON_FASTI2C_EN        (0x3 << 8)
/*!< I2C Fast-mode Plus and high-speed slave */
#define IOCON_INPFILT_OFF       (0x1 << 8)
/*!< Input filter Off for GPIO pins */
#define IOCON_INPFILT_ON        (0x0 << 8)
/*!< Input filter On for GPIO pins */
#define IOCON_OPENDRAIN_EN      (0x1 << 10)
/*!< Enables open-drain function */
#define IOCON_S_MODE_0CLK       (0x0 << 11)
/*!< Bypass input filter */
#define IOCON_S_MODE_1CLK       (0x1 << 11)
/*!< Input pulses shorter than 1 filter clock are rejected */
#define IOCON_S_MODE_2CLK       (0x2 << 11)
/*!< Input pulses shorter than 2 filter clock2 are rejected */
#define IOCON_S_MODE_3CLK       (0x3 << 11)
/*!< Input pulses shorter than 3 filter clock2 are rejected */
#define IOCON_S_MODE(clks)      ((clks) << 11)
/*!< Select clocks for digital input filter mode */
#define IOCON_CLKDIV(div)       ((div) << 13)
/*!< Select peripheral clock divider for input filter sampling clock, 2^n, n = 0 - 6 */
#define LPC_IOCON               ((LPC_IOCON_T     *) LPC_IOCON_BASE)
#define LPC_GPIO                ((LPC_GPIO_T      *) LPC_GPIO_PORT_BASE)
STATIC const PINMUX_GRP_T pinmuxing[] = {
    {0, 25, (IOCON_FUNC0 | IOCON_MODE_PULLDOWN | IOCON_DIGITAL_EN)},
    {0, 26, (IOCON_FUNC0 | IOCON_MODE_PULLDOWN | IOCON_DIGITAL_EN)},
    {0, 27, (IOCON_FUNC0 | IOCON_MODE_PULLDOWN | IOCON_DIGITAL_EN)},
};
    __STATIC_INLINE void Chip_IOCON_PinMuxSet(LPC_IOCON_T * pIOCON, uint8_t port, uint8_t pin, uint32_t modefunc)
```

```c
{
    pIOCON-> PIO[port][pin] = modefunc;
}

__STATIC_INLINE void Chip_GPIO_WritePortBit(LPC_GPIO_T * pGPIO, uint32_t port, uint8_t pin, bool setting)
{
    pGPIO-> B[port][pin] = setting;
}

/* Set all I/O Control pin muxing */
void Chip_IOCON_SetPinMuxing(LPC_IOCON_T * pIOCON, const PINMUX_GRP_T * pinArray, uint32_t arrayLength)
{
    uint32_t ix;

    for (ix = 0; ix < arrayLength; ix++) {
        Chip_IOCON_PinMuxSet(pIOCON, pinArray[ix].port, pinArray[ix].pin, pinArray[ix].modefunc);
    }
}

/* Set Direction for a GPIO port */
void Chip_GPIO_SetDir(LPC_GPIO_T * pGPIO, uint8_t portNum, uint32_t bitValuet)
{
    pGPIO -> DIR[portNum] ^= bitValue;
}
#define PORT_PIN_MST 0
#define PIN_AIN1_BIN1 25
#define PIN_AIN2_BIN2 26
#define PIN_NSLEEP 27
//参考程序入口
Chip_IOCON_SetPinMuxing(LPC_IOCON, pinmuxing, sizeof(pinmuxing)/sizeof(PINMUX_GRP_T));
//初始化 LPC5410，让 P0_25、P0_26 和 P0_27 三个引脚配置成 GPIO 的属性
Chip_GPIO_WritePortBit(LPC_GPIO, PORT_PIN_MST, PIN_NSLEEP,1);
//让 DRV8833 进入正常的工作模式
Chip_GPIO_WritePortBit(LPC_GPIO, PORT_PIN_MST, PIN_AIN1_BIN1,1);
Chip_GPIO_WritePortBit(LPC_GPIO, PORT_PIN_MST, PIN_AIN2_BIN2,0);
```

```
//先配置P0_25、P0_26 为输出 AIN1&BIN1 和 AIN2 一高一低
usleep_mS(1);
//然后开始发送数据
//Sent 1,两次取反
Chip_GPIO_SetDir(LPC_GPIO, PORT_PIN_MST, 0x06000000);
//AIN1&BIN1 取反,AIN2&BIN2 取反
usleepy_uS(200);
Chip_GPIO_SetDir(LPC_GPIO, PORT_PIN_MST, 0x06000000);
//AIN1&BIN1 取反,AIN2&BIN2 取反
usleep_uS(200);

//Sent 0,取反保持
Chip_GPIO_SetDir(LPC_GPIO, PORT_PIN_MST, 0x06000000);
//AIN1&BIN1 取反,AIN2&BIN2 取反
usleep_uS(400);

//Sent 1,两次取反
Chip_GPIO_SetDir(LPC_GPIO, PORT_PIN_MST, 0x06000000);
//AIN1&BIN1 取反,AIN2&BIN2 取反
usleep_uS(200);
Chip_GPIO_SetDir(LPC_GPIO, PORT_PIN_MST, 0x06000000);
//AIN1&BIN1 取反,AIN2&BIN2 取反
usleep_uS(200);

Chip_GPIO_SetDir(LPC_GPIO, PORT_PIN_MST, 0x06000000);
//AIN1&BIN1 取反,AIN2&BIN2 取反
usleep_uS(200);
```

如上所述为在嵌入式平台中的应用案例说明,在有些 Android 移动端的平台中 AIN1&BIN1 和 AIN2&BIN2 的引脚会连接到主机端支持安全访问的区域,这样至少可以在 kernel 层面上避免当系统获取 ROOT 权限后,被程序发送的恶意逻辑控制电平所控制,从而起到安全隔离防护的作用。

4.3 MST 技术与主流近场支付方案对比

目前移动智能设备端支持的线下主流的两种支付方式有基于 QR 条码方式和基于 NFC 方式。前者主要借助于摄像头和显示屏幕,即用户终端应用软件生成订单

号的QR条码,收单终端通过摄像头扫描QR条码,然后通过系统软件把数据解析出来(需要支付的用户信息和订单号等),借助后台系统实现清结算,或者用户终端开启支付应用软件,通过摄像头去扫描商家提供的QR条码,输入金额后完成支付过程;后者基于NFC的支付方式,主要有三种,分别为基于SWP SIM卡方案、嵌入式安全单元eSE方式和HCE主机端软件模拟方式。

QR条码的优势为对于用户端的智能支付终端改造极小,例如对于智能手机来讲摄像头和屏幕这是一个最基本的硬件配置,也就是说只需要安装和更新应用程序就可以完成QR条码支付功能,但是在商户收单系统端则需要进行QR条码扫描和后台系统对接的软硬件的系统升级方可支持。基于NFC方式的SWP SIM、eSE和HCE的前提条件就是用户端的支付设备一定是已经集成了NFC软硬件系统,NFC主要完成支付设备和收单设备之间的一个非接触的通信过程,具体交易支付的部分由SWP SIM、eSE或者HCE软件模拟方式完成。表4-12为基于QR条码、NFC和MST的参考数据对比。

表4-12 QR条码、NFC和MST的参数对比一览表

项 目	QR条码	NFC	MST
通信方式	摄像头	射频场	磁场
用户终端现状	全部的智能手机 (配置有摄像头)	2016年后有近50%的 智能手机支持	三星公司的部 分智能手机
POS终端现状	中国市场普及率极高 (配置有摄像头或者屏幕)	2014年统计数据显示全球有 近15%左右的POS机支持	支持磁条卡 的POS机
应用案例	AliPay,Wechat Pay	Apple Pay, Huawei Pay, Samsung Pay, Mi Pay	Samsung Pay

用户终端现状和POS终端的数据来源于EY(安永)咨询机构。

5 基于Android系统的NFC实现框架

从2010年Android系统默认集成NFC协议栈以来，NFC技术本身得到了巨大的发展，其主要原因有两个：第一，Android系统属于全开源系统，吸引了大量的行业资源和从业者，另外Andriod系统本身主要聚焦在定义框架方面，对于协议栈之类的代码广泛接受供应商的开源贡献，例如像NFC协议栈的中间件代码目前主要集成的供应商有博通和恩智浦两家公司。第二，Android设计框架中把目前主流的SE安全单元的几种典型应用都囊括其中，换句话说就是如果以设备制造商为SE安全单元的主体，那么就可以使用嵌入式的安全单元eSE方式；如果网络运营商希望运营SE安全单元，那么可以使用SWP SIM方案，或者通过在原有的SD卡上面增加一个特殊的SWP接口，支持SE安全单元嵌入到SD方案中进行方案运营，还有就是支持以通过主机端软件模拟HCE方式的方案运营。因为整个系统的框架有了扩充性，设备制造方、网络运营商、银行系统以及第三方的开发者都有了不同的接入接口和实现方式，再加上整个系统的源代码全部是开源的，所以这些年NFC技术发展突飞猛进。

目前，在Google的AOSP（Android Open-Source Project）服务器上依然能看到external文件夹下面保留有一个名叫libnfc-nxp的文件夹，其就是为了适配之前市面上原有的一些主机端与NFC控制器之间以HCI为标准接口的控制器设备，再后来第二代以后的NFC射频控制器前端芯片都采用了NFC论坛定义的NCI统一标准接口，直接参考libnfc-nci文件夹中的代码即可。规范ETSI TS 102 622主要定义了NFC的HCI标准，这个标准中有2010年底第一代商用NFC射频控制器前端芯片与主机端之间的通信接口，因为是当初NFC论坛还没有定义NFC射频控制器前端芯片与主机端的NCI标准，所以那一代的NFC芯片只是在原始的HCI数据包的基础上做了一些扩展，以及在数据格式上增加了数据长度头和使用了类似SWP定义的结尾CRC校验码等。但目前HCI的主要应用方向还是在SE安全芯片与非接触射频前端链接方面，这里的SE安全芯片的封装载体包括内嵌式和SWP-SIM等，所以在本章接下来的篇幅并不会介绍HCI部分。这里所说的主机端与NFC控制器

5 基于 Android 系统的 NFC 实现框架

前端的接口协议全部使用 NCI 接口协议。

对于基于 Android 系统的 NFC 应用开发者而言,需要重点关注 Android 提供的与 NFC 相关的 API 开发接口,而这些 API 接口主要是提供前面曾经介绍过的三种 NFC 的功能。Android 系统中的 NFC 功能主要是通过后台以服务的形式提供给应用层的。这种用法有点类似于我们平时使用的 USB 接口设备,只要 NFC 功能已经打开,当有外部的 POS 终端靠近设备或者扫描到外部的 NFC 卡片和标签时,NFC 的服务程序以及协议栈代码就会迅速地把一些最基本的底层协议握手程序跑完,然后把应用接口的数据反馈到上层应用,再由之前已经注册的 NFC 应用程序去处理之后的应用数据,这样就可以实现多个上层的 NFC 应用程序共享这个服务。如果正在上层应用数据通信时 NFC 服务进程出现异常或者重启,那么此刻上层的应用也会受到波及。

这里以一个 NFC 的标签应用程序为例进行介绍。首先需要注册相关的 intent 到 AndroidManifest.xml 文件中,这样当 Android 设备侦测到外部的 NFC 卡片或者有标签靠近时,NFC Service 会唤醒之前注册过的 intent 列表,在这里就是唤醒 NFC Tag 应用程序;然后 NFC Tag 应用程序通过调用 API 接口,到 NFC 协议栈中去解析和处理上层应用与 NFC 射频前端控制器之间的交互数据(其中 NFC 协议栈中的 HAL 层以及下面的驱动层与实际的 NFC 硬件方案和电路相关),再到最下面控制实际的 NFC 硬件电路中通过射频场的方式把外部的卡片或者标签数据读回来或者写进去。所以 Android 系统中一个最简化的 NFC 软件系统可以分成三层:最上层为标准的应用程序开发接口,中间层为实现 NFC 功能的协议层,最下层为 NFC 驱动程序。如图 5-1 所示为 Android 系统中一个没有包含 SE 安全单元接口的最基本的代码构成层次。

1. API 应用程序接口

该接口主要供上层应用程序开发使用,上层应用程序的 API 和 NFC Service 服务之间通过 Android 的 IPC (Inter-Process Communication)进程间通信机制进行数据交换。

2. NFC 中间层

(1) NFC Service

NFC Service 为 Android 系统中调用 NFC 模块的初始化入口。在 NFC 应用初

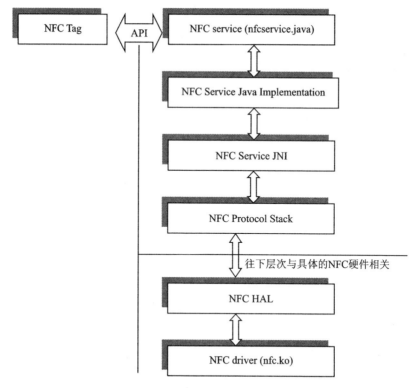

图 5-1 Android 系统 NFC 的流程框图

期阶段,采用内嵌安全单元模块 eSE 方式;有些应用场景,主机端与安全单元 SE 进行数据交换的通道也是采用 NFC Service 的方式。这样在这个服务程序中就不只有 NFC 射频前端控制器的实现部分,还会包含主机端与 SE 之间 APDU 数据交互的 OPEN、CLOSE、TRANSCEIVE 等接口函数。

(2) NFC Service Java Implementation

NFC Service Java Implementation 为上层应用程序调用 Java 接口的 API 函数实体的实现,对应 Framework 层相关的接口。

(3) NFC Service JNI

NFC Service JNI 为 Java 层与 NFC 协议栈 C/C++实现的代码连接的接口层。Android 上层的 Application 和 Framework 都是使用 Java 编写的,但底层代码包括系统和使用众多的库文件都是使用 C/C++编写的,所以上层 Java 要调用底层 C/C++编写的函数或者库文件,必须通过 Android 定义的 Java JNI 标准来实现。

(4) NFC Protocol Stack

NFC Protocol Stack 为 NFC 的相关协议层的代码实现,如前面介绍过的 NCI,

NDEF 等协议实现,其主要原理为主机端通过 NCI 标准的指令对 NFC 射频前端控制器实现和控制它的三种 NFC 模式。

(5) NFC HAL

NFC HAL(Hardware Abstraction Layer)为硬件抽象层。其目的是使系统框架更具灵活性,方便适配不同 NFC 硬件供应商之间的私有命令和协议等,例如大部分的 NFC 控制器中的固件升级指令和协议都是采用各自定义的私有协议等;另外,对于实现的 NCI 状态机的方式可能也是会有少许差别,有了这个层次的设计后,不同的 NFC 组件供应商就可以自己方便地把差异化的东西修改到这个层次中,而不用去改动别的代码模块。

3. NFC 设备驱动程序

该驱动程序符合 Linux 内核驱动程序的封装接口,为上层 NFC 协议栈提供接口,但是屏蔽了底层的实现细节。驱动程序以节点的形式存在文件系统中,驱动程序中与底层硬件打交道的主要有 READ、WRITE 和 IOCTL 三个函数,另外还有支持 Android 系统调用的函数,例如 INIT、EXIT、OPEN、CLOSE、PROBE 和 REMOVE 等。驱动程序的实现实际上就是结合具体的硬件电路部分来编写的,目前主流的主机端与 NFC 射频控制前端连接的接口有 I^2C、UART 和 SPI 等,另外还包含例如硬件中断、复位引脚、时钟和电源管理的物理连接引脚等,一起构成驱动程序。如图 5-2 所示为 NFC 驱动程序的框架,它符合标准的 Linxu 内核驱动程序的设计框架。

现在一般 Android 都启用了 SE-Linux 的机制,在调试 NFC 驱动节点权限时,当启动了 SE-Linux 的默认安全等级从 permissive(宽容模式)升到了 enforcing(强制模式)时,之前在/dev/pn* 这个节点的权限也需要对应更改 sepolicy 文件,否则访问时会被拒绝,代码如下:

```
W/com.android.nfc(5936): type = 1400 audit(0.0:115): avc: denied { read write } for
comm = 4173796E635461736B202335 name = "pn547" dev = "tmpfs" ino = 11384
scontext = u:r:nfc:s0 tcontext = u:object_r:device:s0 tclass = chr_file permissive = 0
D/NxpTml (2046): Opening port = /dev/pn547
E/NxpTml (2046): _i2c_open() Failed: retval ffffffff
```

如果不想采用 enforcing 模式或为了调试方便,那么可以采用如下的方法:

> 通过"adb shell su - c setenforce 0"命令或者把 setenforce 0 加入到 init.rc 中禁用 enforcing 模式即可。

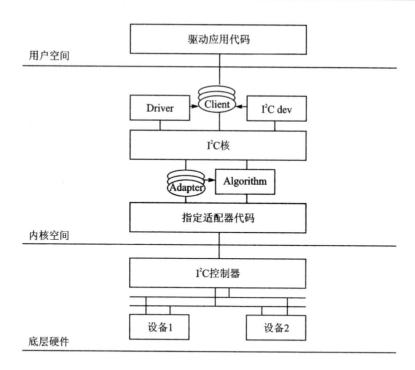

图 5-2　NFC 驱动程序的框架

➢ 直接更改 /etc/selinux/config 里面的 SE-Linux 设置，把 SELINUX=enforcing，改成 SELINUX=permissive or SELINUX=disable，这样退出 enforcing 模式即可。

如果想用 enforcing 模式，那么需要做如下修改：

修改 sepolicy 文件（可能会在 device/qcom/sepolicy/common 或者 external/sepolicy 目录下）。

➢ 在 file_contexts 文件中添加"/dev/pn547　u:object_r:nfc_device:s0"。

➢ 在 app.te 文件中添加"neverallow { appdomain -nfc} nfc_device:chr_file { read write}"。

➢ 在 device.te 文件中添加"type nfc_device, dev_type"。

➢ 在 nfc.te 文件中添加"allow nfc nfc_device:chr_file rw_file_perms"。

以恩智浦的 PN553 射频控制器为例，在高通平台上适配的一个参考的 I²C 接口的驱动代码如下。

头文件 PN553.h 参考如下：

```
/*
 * Copyright (C) 2010 Trusted Logic S.A.
 *
 * This program is free software; you can redistribute it and/or modify
 * it under the terms of the GNU General Public License as published by
 * the Free Software Foundation; either version 2 of the License, or
 * (at your option) any later version.
 *
 * This program is distributed in the hope that it will be useful,
 * but WITHOUT ANY WARRANTY; without even the implied warranty of
 * MERCHANTABILITY or FITNESS FOR A PARTICULAR PURPOSE.  See the
 * GNU General Public License for more details.
 *
 * You should have received a copy of the GNU General Public License
 * along with this program; if not, write to the Free Software
 * Foundation, Inc., 59 Temple Place, Suite 330, Boston, MA 02111-1307 USA
 */
/******************************************************************
 *
 *  The original Work has been changed by NXP Semiconductors.
 *
 *  Copyright (C) 2013-2014 NXP Semiconductors
 *  *
 * This program is free software; you can redistribute it and/or modify
 * it under the terms of the GNU General Public License as published by
 * the Free Software Foundation; either version 2 of the License, or
 * (at your option) any later version.
 *
 * This program is distributed in the hope that it will be useful,
 * but WITHOUT ANY WARRANTY; without even the implied warranty of
 * MERCHANTABILITY or FITNESS FOR A PARTICULAR PURPOSE.  See the
 * GNU General Public License for more details.
 *
 * You should have received a copy of the GNU General Public License
 * along with this program; if not, write to the Free Software
 * Foundation, Inc., 59 Temple Place, Suite 330, Boston, MA 02111-1307 USA
 *
 ******************************************************************/
```

```c
#ifndef _PN553_H_
#define _PN553_H_
#define PN544_MAGIC 0xE9

/*
 * PN544 power control via ioctl
 * PN544_SET_PWR(0): power off
 * PN544_SET_PWR(1): power on
 * PN544_SET_PWR(2): reset and power on with firmware download enabled
 */
#define PN544_SET_PWR    _IOW(PN544_MAGIC, 0x01, long)

/*
 * SPI Request NFCC to enable p61 power, only in param
 * Only for SPI
 * level 1 = Enable power
 * level 0 = Disable power
 */
#define P61_SET_SPI_PWR    _IOW(PN544_MAGIC, 0x02, long)

/* SPI or DWP can call this ioctl to get the current
 * power state of P61
 *
 */
#define P61_GET_PWR_STATUS    _IOR(PN544_MAGIC, 0x03, long)

/* DWP side this ioctl will be called
 * level 1 = Wired access is enabled/ongoing
 * level 0 = Wired access is disalbed/stopped
 */
#define P61_SET_WIRED_ACCESS _IOW(PN544_MAGIC, 0x04, long)

/*
  NFC Init will call the ioctl to register the PID with the i2c driver
 */
#define P544_SET_NFC_SERVICE_PID _IOW(PN544_MAGIC, 0x05, long)

/*
```

5 基于 Android 系统的 NFC 实现框架

```
   NFC and SPI will call the ioctl to get the i2c/spi bus access
*/
#define P544_GET_ESE_ACCESS _IOW(PN544_MAGIC, 0x06, long)
/*
   NFC and SPI will call the ioctl to update the power scheme
*/
#define P544_SET_POWER_SCHEME _IOW(PN544_MAGIC, 0x07, long)

/*
   NFC will call the ioctl to release the svdd protection
*/
#define P544_REL_SVDD_WAIT _IOW(PN544_MAGIC, 0x08, long)


/* SPI or DWP can call this ioctl to get the current
 * power state of P61
 *
*/
#define PN544_SET_DWNLD_STATUS_IOW(PN544_MAGIC, 0x09, long)
/*
   NFC will call the ioctl to release the dwp on/off protection
*/
#define P544_REL_DWPONOFF_WAIT _IOW(PN544_MAGIC, 0x0A, long)

/*
   NFC will call the ioctl to start Secure Timer
*/

#define P544_SECURE_TIMER_SESSION _IOW(PN544_MAGIC, 0x0B, long)


#define MAX_ESE_ACCESS_TIME_OUT_MS 200 /* 100 milliseconds */

typedef enum p61_access_state{
    P61_STATE_INVALID = 0x0000,
    P61_STATE_IDLE = 0x0100, /* p61 is free to use */
    P61_STATE_WIRED = 0x0200,   /* p61 is being accessed by DWP (NFCC) */
    P61_STATE_SPI = 0x0400, /* P61 is being accessed by SPI */
    P61_STATE_DWNLD = 0x0800, /* NFCC fw download is in progress */
    P61_STATE_SPI_PRIO = 0x1000, /* Start of p61 access by SPI on priority */
```

```c
    P61_STATE_SPI_PRIO_END = 0x2000, /* End of p61 access by SPI on priority */
    P61_STATE_SPI_END = 0x4000,
    P61_STATE_JCP_DWNLD = 0x8000,/* JCOP downlad in progress */
    P61_STATE_SECURE_MODE = 0x100000, /* secure mode state */
    P61_STATE_SPI_SVDD_SYNC_START = 0x0001, /* ESE_VDD Low req by SPI */
    P61_STATE_SPI_SVDD_SYNC_END = 0x0002, /* ESE_VDD is Low by SPI */
    P61_STATE_DWP_SVDD_SYNC_START = 0x0004, /* ESE_VDD Low req by Nfc */
    P61_STATE_DWP_SVDD_SYNC_END = 0x0008 /* ESE_VDD is Low by Nfc */
}p61_access_state_t;

typedef enum chip_type_pwr_scheme{
    PN67T_PWR_SCHEME = 0x01,
    PN80T_LEGACY_PWR_SCHEME,
    PN80T_EXT_PMU_SCHEME,
}chip_pwr_scheme_t;

typedef enum jcop_dwnld_state{
    JCP_DWNLD_IDLE = P61_STATE_JCP_DWNLD,   /* jcop dwnld is ongoing */
    JCP_DWNLD_INIT = 0x8010,                /* jcop dwonload init state */
    JCP_DWNLD_START = 0x8020,               /* download started */
    JCP_SPI_DWNLD_COMPLETE = 0x8040,        /* jcop download complete in spi interface */
    JCP_DWP_DWNLD_COMPLETE = 0x8080,        /* jcop download complete */
} jcop_dwnld_state_t;

struct pn544_i2c_platform_data {
    unsigned int irq_gpio;
    unsigned int ven_gpio;
    unsigned int firm_gpio;
    unsigned int ese_pwr_gpio; /* gpio to give power to p61, only TEE should use this */
    unsigned int iso_rst_gpio; /* gpio used for ISO hard reset P73 */
};

struct hw_type_info {
    /*
     * Response of get_version_cmd will be stored in data
     * byte structure :
     * byte 0 - 1    : Header
     * byte 2        : Status
```

```
    *    byte 3              : Hardware Version
    *    byte 4              : ROM code
    *    byte 5              : 0x00 constant
    *    byte 6 - 7          : Protected data version
    *    byte 8 - 9          : Trim data version
    *    byte 10 - 11        : FW version
    *    byte 12 - 13        : CRC
    *    */
    char data[20];
    int len;
};
#endif
```

源文件 PN553.c 参考如下:

```
/*
 * Copyright (C) 2010 Trusted Logic S.A.
 *
 * This program is free software; you can redistribute it and/or modify
 * it under the terms of the GNU General Public License as published by
 * the Free Software Foundation; either version 2 of the License, or
 * (at your option) any later version.
 *
 * This program is distributed in the hope that it will be useful,
 * but WITHOUT ANY WARRANTY; without even the implied warranty of
 * MERCHANTABILITY or FITNESS FOR A PARTICULAR PURPOSE. See the
 * GNU General Public License for more details.
 *
 * You should have received a copy of the GNU General Public License
 * along with this program; if not, write to the Free Software
 * Foundation, Inc., 51 Franklin St, Fifth Floor, Boston, MA 02110 - 1301 USA
 *
 */
/******************************************************************
 *
 *  The original Work has been changed by NXP Semiconductors.
 *
 *  Copyright (C) 2013 - 2014 NXP Semiconductors
 *
 *  *
```

```
* This program is free software; you can redistribute it and/or modify
* it under the terms of the GNU General Public License as published by
* the Free Software Foundation; either version 2 of the License, or
* (at your option) any later version.
*
* This program is distributed in the hope that it will be useful,
* but WITHOUT ANY WARRANTY; without even the implied warranty of
* MERCHANTABILITY or FITNESS FOR A PARTICULAR PURPOSE. See the
* GNU General Public License for more details.
*
* You should have received a copy of the GNU General Public License
* along with this program; if not, write to the Free Software
* Foundation, Inc., 59 Temple Place, Suite 330, Boston, MA 02111-1307 USA
*
**************************************************************/

#include < linux/kernel.h >
#include < linux/module.h >
#include < linux/fs.h >
#include < linux/slab.h >
#include < linux/init.h >
#include < linux/list.h >
#include < linux/i2c.h >
#include < linux/irq.h >
#include < linux/jiffies.h >
#include < linux/uaccess.h >
#include < linux/delay.h >
#include < linux/interrupt.h >
#include < linux/io.h >
#include < linux/platform_device.h >
#include < linux/gpio.h >
#include < linux/of_gpio.h >
#include < linux/miscdevice.h >
#include < linux/spinlock.h >
#include < asm/siginfo.h >
#include < linux/rcupdate.h >
#include < linux/sched.h >
#include < linux/signal.h >
```

5 基于 Android 系统的 NFC 实现框架

```c
#include <linux/workqueue.h>
/* HiKey Compilation fix */
#define HiKey_620_COMPILATION_FIX 1
#ifndef HiKey_620_COMPILATION_FIX
#include <linux/wakelock.h>
#endif

#include <linux/timer.h>
#include "pn553.h"

#define NEXUS5x     0
#define HWINFO      0
#if NEXUS5x
#undef ISO_RST
#else
#define ISO_RST
#endif
#define DRAGON_NFC 1
#define SIG_NFC 44
#define MAX_BUFFER_SIZE 512
#define MAX_SECURE_SESSIONS 1
/* Macro added to disable SVDD power toggling */
/* #define JCOP_4X_VALIDATION */

struct pn544_dev {
    wait_queue_head_t read_wq;
    struct mutex read_mutex;
    struct i2c_client *client;
    struct miscdevice pn544_device;
    unsigned int ven_gpio;
    unsigned int firm_gpio;
    unsigned int irq_gpio;
    unsigned int ese_pwr_gpio; /* gpio used by SPI to provide power to p61 via NFCC */
#ifdef ISO_RST
    unsigned int iso_rst_gpio; /* ISO-RST pin gpio */
#endif
    struct mutex p61_state_mutex; /* used to make p61_current_state flag secure */
    p61_access_state_t p61_current_state; /* stores the current P61 state */
```

```c
        bool nfc_ven_enabled; /* stores the VEN pin state powered by Nfc */
        bool spi_ven_enabled; /* stores the VEN pin state powered by Spi */
        bool irq_enabled;
        spinlock_t irq_enabled_lock;
        long nfc_service_pid; /* used to signal the nfc the nfc service */
        chip_pwr_scheme_t chip_pwr_scheme;
        unsigned int secure_timer_cnt;
        struct workqueue_struct * pSecureTimerCbWq;
        struct work_struct wq_task;
};
/* HiKey Compilation fix */
#ifndef HiKey_620_COMPILATION_FIX
struct wake_lock nfc_wake_lock;
#if HWINFO
struct hw_type_info hw_info;
#endif
static bool sIsWakeLocked = false;
#endif
static struct pn544_dev * pn544_dev;
static struct semaphore ese_access_sema;
static struct semaphore svdd_sync_onoff_sema;
static struct completion dwp_onoff_sema;
static struct timer_list secure_timer;
static void release_ese_lock(p61_access_state_t p61_current_state);
int get_ese_lock(p61_access_state_t p61_current_state, int timeout);
static long set_jcop_download_state(unsigned long arg);
static long start_seccure_timer(unsigned long timer_value);
static long secure_timer_operation(struct pn544_dev * pn544_dev, unsigned long arg);
#if HWINFO
static void check_hw_info(void);
#endif
#define SECURE_TIMER_WORK_QUEUE "SecTimerCbWq"


static void pn544_disable_irq(struct pn544_dev * pn544_dev)
{
        unsigned long flags;

        spin_lock_irqsave(&pn544_dev -> irq_enabled_lock, flags);
```

```c
        if (pn544_dev -> irq_enabled) {
            disable_irq_nosync(pn544_dev -> client -> irq);
            disable_irq_wake(pn544_dev -> client -> irq);
            pn544_dev -> irq_enabled = false;
        }
        spin_unlock_irqrestore(&pn544_dev -> irq_enabled_lock, flags);
}

static irqreturn_t pn544_dev_irq_handler(int irq, void * dev_id)
{
    struct pn544_dev * pn544_dev = dev_id;

    pn544_disable_irq(pn544_dev);
    /* HiKey Compilation fix */
    #ifndef HiKey_620_COMPILATION_FIX
    if (sIsWakeLocked == false)
    {
        wake_lock(&nfc_wake_lock);
        sIsWakeLocked = true;
    } else {
            pr_debug("%s already wake locked!\n", __func__);
    }
    #endif
    /* Wake up waiting readers */
    wake_up(&pn544_dev -> read_wq);

    return IRQ_HANDLED;
}

static ssize_t pn544_dev_read(struct file * filp, char __user * buf,
                              size_t count, loff_t * offset)
{
    struct pn544_dev * pn544_dev = filp -> private_data;
    char tmp[MAX_BUFFER_SIZE];
    int ret;

    if (count > MAX_BUFFER_SIZE)
        count = MAX_BUFFER_SIZE;
```

```c
//pr_debug("%s : reading %zu bytes.\n", __func__, count);

mutex_lock(&pn544_dev->read_mutex);

if (!gpio_get_value(pn544_dev->irq_gpio)) {
    if (filp->f_flags & O_NONBLOCK) {
        ret = -EAGAIN;
        goto fail;
    }

    while (1) {
        pn544_dev->irq_enabled = true;
        enable_irq(pn544_dev->client->irq);
        enable_irq_wake(pn544_dev->client->irq);
        ret = wait_event_interruptible(
            pn544_dev->read_wq,
            !pn544_dev->irq_enabled);

        pn544_disable_irq(pn544_dev);

        if (ret)
            goto fail;

        if (gpio_get_value(pn544_dev->irq_gpio))
            break;

        pr_warning("%s: spurious interrupt detected\n", __func__);
    }
}

/* Read data */
ret = i2c_master_recv(pn544_dev->client, tmp, count);
#ifndef HiKey_620_COMPILATION_FIX
/* HiKey Compilation fix */
if (sIsWakeLocked == true) {
    wake_unlock(&nfc_wake_lock);
    sIsWakeLocked = false;
}
```

```c
    #endif
        mutex_unlock(&pn544_dev->read_mutex);

        /* pn544 seems to be slow in handling I2C read requests
         * so add 1ms delay after recv operation */
    #if !NEXUS5x
        udelay(1000);
    #endif

        if (ret < 0) {
            pr_err("%s: i2c_master_recv returned %d\n", __func__, ret);
            return ret;
        }
        if (ret > count) {
            pr_err("%s: received too many bytes from i2c (%d)\n", __func__, ret);
            return -EIO;
        }
        if (copy_to_user(buf, tmp, ret)) {
            pr_warning("%s: failed to copy to user space\n", __func__);
            return -EFAULT;
        }
        return ret;

    fail:
        mutex_unlock(&pn544_dev->read_mutex);
        return ret;
}


static ssize_t pn544_dev_write(struct file *filp, const char __user *buf,
                        size_t count, loff_t *offset)
{
    struct pn544_dev *pn544_dev;
    char tmp[MAX_BUFFER_SIZE];
    int ret;

    pn544_dev = filp->private_data;

    if (count > MAX_BUFFER_SIZE)
```

```
            count = MAX_BUFFER_SIZE;

        if (copy_from_user(tmp, buf, count)) {
            pr_err("%s : failed to copy from user space\n", __func__);
            return -EFAULT;
        }

        //pr_debug("%s : writing %zu bytes.\n", __func__, count);
        /* Write data */
        ret = i2c_master_send(pn544_dev->client, tmp, count);
        if (ret != count) {
            pr_err("%s : i2c_master_send returned %d\n", __func__, ret);
            ret = -EIO;
        }

        /* pn544 seems to be slow in handling I2C write requests
         * so add 1ms delay after I2C send oparation */
        udelay(1000);

        return ret;
    }

    static void p61_update_access_state(struct pn544_dev * pn544_dev, p61_access_state_t
                                current_state, bool set)
    {
        pr_info("%s: Enter current_state = %x\n", __func__, pn544_dev->p61_current_state);
        if (current_state)
        {
            if(set){
                if(pn544_dev->p61_current_state == P61_STATE_IDLE)
                    pn544_dev->p61_current_state = P61_STATE_INVALID;
                pn544_dev->p61_current_state |= current_state;
            }
            else{
                pn544_dev->p61_current_state ^= current_state;
                if(!pn544_dev->p61_current_state)
                    pn544_dev->p61_current_state = P61_STATE_IDLE;
```

```c
        }
    }
    pr_info("%s: Exit current_state = %x\n", __func__, pn544_dev -> p61_current_state);
}

static void p61_get_access_state(struct pn544_dev * pn544_dev, p61_access_state_t * 
                                 current_state)
{
    if (current_state == NULL) {
        //* current_state = P61_STATE_INVALID;
        pr_err("%s : invalid state of p61_access_state_t current state  \n", __func__);
    } else {
        * current_state = pn544_dev -> p61_current_state;
    }
}

static void p61_access_lock(struct pn544_dev * pn544_dev)
{
    pr_info("%s: Enter\n", __func__);
    mutex_lock(&pn544_dev -> p61_state_mutex);
    pr_info("%s: Exit\n", __func__);
}

static void p61_access_unlock(struct pn544_dev * pn544_dev)
{
    pr_info("%s: Enter\n", __func__);
    mutex_unlock(&pn544_dev -> p61_state_mutex);
    pr_info("%s: Exit\n", __func__);
}


static int signal_handler(p61_access_state_t state, long nfc_pid)
{
    struct siginfo sinfo;
    pid_t pid;
    struct task_struct * task;
    int sigret = 0, ret = 0;
    pr_info("%s: Enter\n", __func__);

    if(nfc_pid == 0)
```

```c
        {
            pr_info("nfc_pid is clear don't call signal_handler.\n");
        }
        else
        {
            memset(&sinfo, 0, sizeof(struct siginfo));
            sinfo.si_signo = SIG_NFC;
            sinfo.si_code = SI_QUEUE;
            sinfo.si_int = state;
            pid = nfc_pid;

            task = pid_task(find_vpid(pid), PIDTYPE_PID);
            if(task)
            {
                pr_info("%s.\n", task->comm);
                sigret = force_sig_info(SIG_NFC, &sinfo, task);
                if(sigret < 0){
                    pr_info("send_sig_info failed..... sigret %d.\n", sigret);
                    ret = -1;
                    //msleep(60);
                }
            }
            else{
                pr_info("finding task from PID failed\r\n");
                ret = -1;
            }
        }
        pr_info("%s: Exit ret = %d\n", __func__, ret);
        return ret;
    }
    static void svdd_sync_onoff(long nfc_service_pid, p61_access_state_t origin)
    {
        int timeout = 100; //100 ms timeout
        unsigned long tempJ = msecs_to_jiffies(timeout);
        pr_info("%s: Enter nfc_service_pid: %ld\n", __func__, nfc_service_pid);
        if(nfc_service_pid)
        {
            if (0 == signal_handler(origin, nfc_service_pid))
```

```c
        {
            sema_init(&svdd_sync_onoff_sema, 0);
            pr_info("Waiting for svdd protection response");
            if(down_timeout(&svdd_sync_onoff_sema, tempJ) != 0)
            {
                pr_info("svdd wait protection: Timeout");
            }
            pr_info("svdd wait protection : released");
        }
    }
    pr_info("%s: Exit\n", __func__);
}
static int release_svdd_wait(void)
{
    pr_info("%s: Enter \n", __func__);
    up(&svdd_sync_onoff_sema);
    pr_info("%s: Exit\n", __func__);
    return 0;
}


static void dwp_OnOff(long nfc_service_pid, p61_access_state_t origin)
{
    int timeout = 100; //100 ms timeout
    unsigned long tempJ = msecs_to_jiffies(timeout);
    if(nfc_service_pid)
    {
        if (0 == signal_handler(origin, nfc_service_pid))
        {
            init_completion(&dwp_onoff_sema);
            if(wait_for_completion_timeout(&dwp_onoff_sema, tempJ) != 0)
            {
                pr_info("Dwp On/off wait protection: Timeout");
            }
            pr_info("Dwp On/Off wait protection : released");
        }
    }
}
static int release_dwpOnOff_wait(void)
```

```
{
    pr_info("%s: Enter \n", __func__);
    complete(&dwp_onoff_sema);
    return 0;
}


static int pn544_dev_open(struct inode * inode, struct file * filp)
{
    struct pn544_dev * pn544_dev = container_of(filp -> private_data,
            struct pn544_dev,
            pn544_device);

    filp -> private_data = pn544_dev;

    pr_debug("%s : %d, %d\n", __func__, imajor(inode), iminor(inode));

    return 0;
}


static int set_nfc_pid(unsigned long arg)
{
    pr_info("%s : The NFC Service PID is %ld\n", __func__, arg);
    pn544_dev -> nfc_service_pid = arg;
    return 0;
}


long pn544_dev_ioctl(struct file * filp, unsigned int cmd, unsigned long arg)
{
    pr_info("%s :enter cmd = %u, arg = %ld\n", __func__, cmd, arg);

    /* Free pass autobahn area, not protected. Use it carefullly. START */
    switch(cmd)
    {
        case P544_GET_ESE_ACCESS:
            return get_ese_lock(P61_STATE_WIRED, arg);
        break;
        case P544_REL_SVDD_WAIT:
            return release_svdd_wait();
```

```
            break;
        case P544_SET_NFC_SERVICE_PID:
            return set_nfc_pid(arg);
        break;
        case P544_REL_DWPONOFF_WAIT:
            return release_dwpOnOff_wait();
        break;
        default:
        break;
    }
    /* Free pass autobahn area, not protected. Use it carefullly. END */

    p61_access_lock(pn544_dev);
    switch (cmd) {
    case PN544_SET_PWR:
    {
        p61_access_state_t current_state = P61_STATE_INVALID;
        p61_get_access_state(pn544_dev, &current_state);
        if (arg == 2) {
            if (current_state & (P61_STATE_SPI|P61_STATE_SPI_PRIO) && (pn544_dev->chip_pwr_scheme != PN80T_EXT_PMU_SCHEME))
            {
                /* NFCC fw/download should not be allowed if p61 is used
                 * by SPI
                 */
                pr_info(" % s NFCC should not be allowed to reset/FW download \n", __func__);
                p61_access_unlock(pn544_dev);
                return -EBUSY; /* Device or resource busy */
            }
            pn544_dev->nfc_ven_enabled = true;
            if ((pn544_dev->spi_ven_enabled == false && !(pn544_dev->secure_timer_cnt))
                || (pn544_dev->chip_pwr_scheme == PN80T_EXT_PMU_SCHEME))
            {
                /* power on with firmware download (requires hw reset)
                 */
                pr_info(" % s power on with firmware\n", __func__);
```

```
                    gpio_set_value(pn544_dev -> ven_gpio, 1);
                    msleep(10);
                    if (pn544_dev -> firm_gpio) {
                        p61_update_access_state(pn544_dev, P61_STATE_DWNLD, true);
                        gpio_set_value(pn544_dev -> firm_gpio, 1);
                    }
                    msleep(10);
                    gpio_set_value(pn544_dev -> ven_gpio, 0);
                    msleep(10);
                    gpio_set_value(pn544_dev -> ven_gpio, 1);
                    msleep(10);
                }
            } else if (arg == 1) {
                /* power on */
                pr_info("%s power on\n", __func__);
                if (pn544_dev -> firm_gpio) {
                    if ((current_state & (P61_STATE_WIRED|P61_STATE_SPI|P61_STATE_SPI_PRIO)) == 0){
                        p61_update_access_state(pn544_dev, P61_STATE_IDLE, true);
                    }
                    if(current_state & P61_STATE_DWNLD){
                        p61_update_access_state(pn544_dev, P61_STATE_DWNLD, false);
                    }
                    gpio_set_value(pn544_dev -> firm_gpio, 0);
                }


                pn544_dev -> nfc_ven_enabled = true;
                if (pn544_dev -> spi_ven_enabled == false || (pn544_dev -> chip_pwr_scheme == PN80T_EXT_PMU_SCHEME)) {
                    gpio_set_value(pn544_dev -> ven_gpio, 1);
                }
            } else if (arg == 0) {
                /* power off */
                pr_info("%s power off\n", __func__);
                if (pn544_dev -> firm_gpio) {
                    if ((current_state & (P61_STATE_WIRED|P61_STATE_SPI|P61_STATE_SPI_PRIO)) == 0){
                        p61_update_access_state(pn544_dev, P61_STATE_IDLE, true);
```

```
            }
            gpio_set_value(pn544_dev->firm_gpio,0);
        }

        pn544_dev->nfc_ven_enabled=false;
        /* Don't change Ven state if spi made it high */
        if((pn544_dev->spi_ven_enabled==false && !(pn544_dev->secure_timer_cnt))
            ||(pn544_dev->chip_pwr_scheme==PN80T_EXT_PMU_SCHEME)){
            gpio_set_value(pn544_dev->ven_gpio,0);
        }
        /* HiKey Compilation fix */
        #ifndef HiKey_620_COMPILATION_FIX
        if(sIsWakeLocked==true){
            wake_unlock(&nfc_wake_lock);
            sIsWakeLocked=false;
        }
        #endif
    } else if(arg==3){
        /* NFC Service called ISO-RST */
        p61_access_state_t current_state=P61_STATE_INVALID;
        p61_get_access_state(pn544_dev,&current_state);
        if(current_state & (P61_STATE_SPI|P61_STATE_SPI_PRIO)){
            p61_access_unlock(pn544_dev);
            return -EPERM; /* Operation not permitted */
        }
        if(current_state & P61_STATE_WIRED){
            p61_update_access_state(pn544_dev,P61_STATE_WIRED,false);
        }
#ifdef ISO_RST
        gpio_set_value(pn544_dev->iso_rst_gpio,0);
        msleep(50);
        gpio_set_value(pn544_dev->iso_rst_gpio,1);
        msleep(50);
        pr_info("%s ISO RESET from DWP DONE\n",__func__);
#endif
    }
    else{
```

```c
                pr_err("%s bad arg %lu\n", __func__, arg);
                /* changed the p61 state to idle */
                p61_access_unlock(pn544_dev);
                return -EINVAL;
            }
        }
        break;
    case P61_SET_SPI_PWR:
    {
        p61_access_state_t current_state = P61_STATE_INVALID;
        p61_get_access_state(pn544_dev, &current_state);
        if (arg == 1) {
            pr_info("%s : PN61_SET_SPI_PWR - power on ese\n", __func__);
            if ((current_state & (P61_STATE_SPI|P61_STATE_SPI_PRIO)) == 0)
            {
                p61_update_access_state(pn544_dev, P61_STATE_SPI, true);
                /* To handle triple mode protection signal
                NFC service when SPI session started */
                if (!(current_state & P61_STATE_JCP_DWNLD)){
                    if(pn544_dev->nfc_service_pid){
                        pr_info("nfc service pid %s  ---- %ld", __func__,
pn544_dev->nfc_service_pid);
                    /* signal_handler(P61_STATE_SPI, pn544_dev->nfc_service_pid); */
                        dwp_OnOff(pn544_dev->nfc_service_pid, P61_STATE_SPI);
                    }
                    else{
                        pr_info(" invalid nfc service pid.... signalling failed%
s   ---- %ld", __func__, pn544_dev->nfc_service_pid);
                    }
                }
                pn544_dev->spi_ven_enabled = true;

                if(pn544_dev->chip_pwr_scheme == PN80T_EXT_PMU_SCHEME)
                    break;

                if (pn544_dev->nfc_ven_enabled == false)
                {
                    /* provide power to NFCC if, NFC service not provided */
```

5 基于 Android 系统的 NFC 实现框架

```
                gpio_set_value(pn544_dev->ven_gpio, 1);
                msleep(10);
            }
            /* pull the gpio to high once NFCC is power on */
            gpio_set_value(pn544_dev->ese_pwr_gpio, 1);

            /* Delay (10ms) after SVDD_PWR_ON to allow JCOP to bootup (5ms jcop boot time + 5ms guard time) */
            usleep_range(10000, 12000);

        } else {
            pr_info("%s : PN61_SET_SPI_PWR - power on ese failed \n", __func__);
            p61_access_unlock(pn544_dev);
            return -EBUSY; /* Device or resource busy */
        }
    } else if (arg == 0) {
        pr_info("%s : PN61_SET_SPI_PWR - power off ese\n", __func__);
        if(current_state & P61_STATE_SPI_PRIO){
            p61_update_access_state(pn544_dev, P61_STATE_SPI_PRIO, false);
            if (!(current_state & P61_STATE_JCP_DWNLD))
            {
                if(pn544_dev->nfc_service_pid){
                    pr_info("nfc service pid %s    ---- %ld", __func__, pn544_dev->nfc_service_pid);
                    if(!(current_state & P61_STATE_WIRED))
                    {
                        svdd_sync_onoff(pn544_dev->nfc_service_pid, P61_STATE_SPI_SVDD_SYNC_START |
                                        P61_STATE_SPI_PRIO_END);
                    }else {
                        signal_handler(P61_STATE_SPI_PRIO_END, pn544_dev->nfc_service_pid);
                    }
                }
                else{
                    pr_info(" invalid nfc service pid....signalling failed %s     ---- %ld", __func__, pn544_dev->nfc_service_pid);
                }
```

```c
            } else if (!(current_state & P61_STATE_WIRED)) {
                svdd_sync_onoff(pn544_dev -> nfc_service_pid, P61_STATE_SPI_SVDD_SYNC_START);
            }
            pn544_dev -> spi_ven_enabled = false;

            if(pn544_dev -> chip_pwr_scheme == PN80T_EXT_PMU_SCHEME)
                break;

            /* if secure timer is running, Delay the SPI close by 25ms after sending End of Apdu to enable eSE go into DPD
               gracefully (20ms after EOS + 5ms DPD settlement time) */
            if(pn544_dev -> secure_timer_cnt)
                usleep_range(25000, 30000);

            if (!(current_state & P61_STATE_WIRED) && !(pn544_dev -> secure_timer_cnt))
            {
#ifndef JCOP_4X_VALIDATION
                gpio_set_value(pn544_dev -> ese_pwr_gpio, 0);
                /* Delay (2.5ms) after SVDD_PWR_OFF for the shutdown settlement time */
                usleep_range(2500, 3000);
#endif
                svdd_sync_onoff(pn544_dev -> nfc_service_pid, P61_STATE_SPI_SVDD_SYNC_END);
            }
#ifndef JCOP_4X_VALIDATION
            if ((pn544_dev -> nfc_ven_enabled == false) && !(pn544_dev -> secure_timer_cnt)) {
                gpio_set_value(pn544_dev -> ven_gpio, 0);
                msleep(10);
            }
#endif
        }else if(current_state & P61_STATE_SPI){
            p61_update_access_state(pn544_dev, P61_STATE_SPI, false);
            if (!(current_state & P61_STATE_WIRED) &&
                (pn544_dev -> chip_pwr_scheme != PN80T_EXT_PMU_SCHEME) &&
```

```c
                        !(current_state & P61_STATE_JCP_DWNLD))
                    {
                        if(pn544_dev -> nfc_service_pid){
                            pr_info("nfc service pid % s    ---- % ld", __func__, pn544_
dev -> nfc_service_pid);
                            svdd_sync_onoff(pn544_dev -> nfc_service_pid, P61_STATE_
SPI_SVDD_SYNC_START | P61_STATE_SPI_END);
                        }
                        else{
                            pr_info(" invalid nfc service pid....signalling failed %
s    ---- % ld", __func__, pn544_dev -> nfc_service_pid);
                        }
                        /* if secure timer is running, Delay the SPI close by 25ms after
sending End of Apdu to enable eSE go into DPD
                        gracefully (20ms after EOS + 5ms DPD settlement time) */
                        if(pn544_dev -> secure_timer_cnt)
                            usleep_range(25000, 30000);

                        if (!(pn544_dev -> secure_timer_cnt)) {
# ifndef JCOP_4X_VALIDATION
                            gpio_set_value(pn544_dev -> ese_pwr_gpio, 0);
                            /* Delay (2.5ms) after SVDD_PWR_OFF for the shutdown set-
tlement time */
                            usleep_range(2500, 3000);
# endif
                            svdd_sync_onoff(pn544_dev -> nfc_service_pid, P61_STATE_
SPI_SVDD_SYNC_END);
                        }
                    }
                    /* If JCOP3.2 or 3.3 for handling triple mode
                    protection signal NFC service */
                    else
                    {
                        if (!(current_state & P61_STATE_JCP_DWNLD))
                        {
                            if(pn544_dev -> nfc_service_pid){
                                pr_info("nfc service pid % s    ---- % ld", __func__,
pn544_dev -> nfc_service_pid);
```

```
                                    if(pn544_dev->chip_pwr_scheme==PN80T_LEGACY_PWR_
SCHEME)
                                    {
                                        svdd_sync_onoff(pn544_dev->nfc_service_pid,
P61_STATE_SPI_SVDD_SYNC_START | P61_STATE_SPI_END);
                                    } else {
                                        signal_handler(P61_STATE_SPI_END, pn544_dev->
nfc_service_pid);
                                    }
                                }
                                else{
                                    pr_info(" invalid nfc service pid.... signalling
failed %s ---- %ld", __func__, pn544_dev->nfc_service_pid);
                                }
                            } else if (pn544_dev->chip_pwr_scheme==PN80T_LEGACY_PWR_
SCHEME) {
                                svdd_sync_onoff(pn544_dev->nfc_service_pid, P61_STATE_
SPI_SVDD_SYNC_START);
                            }
                            if(pn544_dev->chip_pwr_scheme==PN80T_LEGACY_PWR_SCHEME)
                            {
#ifndef JCOP_4X_VALIDATION
                                gpio_set_value(pn544_dev->ese_pwr_gpio, 0);
#endif
                                svdd_sync_onoff(pn544_dev->nfc_service_pid, P61_STATE_
SPI_SVDD_SYNC_END);
                                pr_info("PN80T legacy ese_pwr_gpio off %s", __func__);
                            }
                        }
                        pn544_dev->spi_ven_enabled = false;
                        if (pn544_dev->nfc_ven_enabled == false && (pn544_dev->chip_pwr_
scheme != PN80T_EXT_PMU_SCHEME)
                            && !(pn544_dev->secure_timer_cnt)) {
                            gpio_set_value(pn544_dev->ven_gpio, 0);
                            msleep(10);
                        }
                    } else {
                        pr_err("%s : PN61_SET_SPI_PWR - failed, current_state = %x \n",
```

```
                    __func__, pn544_dev -> p61_current_state);
                p61_access_unlock(pn544_dev);
                return -EPERM; /* Operation not permitted */
            }
        } else if (arg == 2) {
            pr_info("%s : PN61_SET_SPI_PWR - reset\n", __func__);
            if (current_state & (P61_STATE_IDLE | P61_STATE_SPI | P61_STATE_SPI_PRIO)) {
                if (pn544_dev -> spi_ven_enabled == false)
                {
                    pn544_dev -> spi_ven_enabled = true;
                    if ((pn544_dev -> nfc_ven_enabled == false) && (pn544_dev -> chip_pwr_scheme != PN80T_EXT_PMU_SCHEME)) {
                        /* provide power to NFCC if, NFC service not provided */
                        gpio_set_value(pn544_dev -> ven_gpio, 1);
                        msleep(10);
                    }
                }
                if(pn544_dev -> chip_pwr_scheme != PN80T_EXT_PMU_SCHEME && !(pn544_dev -> secure_timer_cnt))
                {
                    svdd_sync_onoff(pn544_dev -> nfc_service_pid, P61_STATE_SPI_SVDD_SYNC_START);
#ifndef JCOP_4X_VALIDATION
                    gpio_set_value(pn544_dev -> ese_pwr_gpio, 0);
#endif
                    svdd_sync_onoff(pn544_dev -> nfc_service_pid, P61_STATE_SPI_SVDD_SYNC_END);
                    msleep(10);
                    if(!gpio_get_value(pn544_dev -> ese_pwr_gpio))
                        gpio_set_value(pn544_dev -> ese_pwr_gpio, 1);
                    msleep(10);
                }
            } else {
                pr_info("%s : PN61_SET_SPI_PWR - reset failed\n", __func__);
                p61_access_unlock(pn544_dev);
                return -EBUSY; /* Device or resource busy */
            }
```

```c
                    }else if(arg == 3){
                        pr_info("%s : PN61_SET_SPI_PWR - Prio Session Start power on ese\n", __func__);
                        if((current_state & (P61_STATE_SPI|P61_STATE_SPI_PRIO)) == 0){
                            p61_update_access_state(pn544_dev, P61_STATE_SPI_PRIO, true);
                            if(current_state & P61_STATE_WIRED){
                                if(pn544_dev->nfc_service_pid){
                                    pr_info("nfc service pid %s   ---- %ld", __func__, pn544_dev->nfc_service_pid);
                                    /*signal_handler(P61_STATE_SPI_PRIO, pn544_dev->nfc_service_pid);*/
                                    dwp_OnOff(pn544_dev->nfc_service_pid, P61_STATE_SPI_PRIO);
                                }
                                else{
                                    pr_info(" invalid nfc service pid....signalling failed%s  ---- %ld", __func__, pn544_dev->nfc_service_pid);
                                }
                            }
                            pn544_dev->spi_ven_enabled = true;
                            if(pn544_dev->chip_pwr_scheme != PN80T_EXT_PMU_SCHEME)
                            {
                                if (pn544_dev->nfc_ven_enabled == false){
                                    /* provide power to NFCC if, NFC service not provided */
                                    gpio_set_value(pn544_dev->ven_gpio, 1);
                                    msleep(10);
                                }
                                /* pull the gpio to high once NFCC is power on */
                                gpio_set_value(pn544_dev->ese_pwr_gpio, 1);

                                /* Delay (10ms) after SVDD_PWR_ON to allow JCOP to bootup (5ms jcop boot time + 5ms guard time) */
                                usleep_range(10000, 12000);
                            }
                        }else {
                            pr_info("%s : Prio Session Start power on ese failed \n", __func__);
                            p61_access_unlock(pn544_dev);
                            return -EBUSY; /* Device or resource busy */
```

```c
            }
        }else if (arg == 4) {
            if (current_state & P61_STATE_SPI_PRIO)
            {
                pr_info("%s : PN61_SET_SPI_PWR - Prio Session Ending...\n", __func__);
                p61_update_access_state(pn544_dev, P61_STATE_SPI_PRIO, false);
                /* after SPI prio timeout, the state is changing from SPI prio to SPI */
                p61_update_access_state(pn544_dev, P61_STATE_SPI, true);
                if (current_state & P61_STATE_WIRED)
                {
                    if(pn544_dev->nfc_service_pid){
                        pr_info("nfc service pid %s    ---- %ld", __func__, pn544_dev->nfc_service_pid);
                        signal_handler(P61_STATE_SPI_PRIO_END, pn544_dev->nfc_service_pid);
                    }
                    else{
                        pr_info(" invalid nfc service pid....signalling failed %s    ---- %ld", __func__, pn544_dev->nfc_service_pid);
                    }
                }
            }
            else
            {
                pr_info("%s : PN61_SET_SPI_PWR -  Prio Session End failed \n", __func__);
                p61_access_unlock(pn544_dev);
                return -EBADRQC; /* Device or resource busy */
            }
        } else if(arg == 5){
            release_ese_lock(P61_STATE_SPI);
        } else if (arg == 6) {
            /* SPI Service called ISO-RST */
            p61_access_state_t current_state = P61_STATE_INVALID;
            p61_get_access_state(pn544_dev, &current_state);
            if(current_state & P61_STATE_WIRED) {
                p61_access_unlock(pn544_dev);
```

```c
                    return - EPERM; /* Operation not permitted */
                }
                if(current_state & P61_STATE_SPI) {
                    p61_update_access_state(pn544_dev, P61_STATE_SPI, false);
                }else if(current_state & P61_STATE_SPI_PRIO) {
                    p61_update_access_state(pn544_dev, P61_STATE_SPI_PRIO, false);
                }
#ifdef ISO_RST
                gpio_set_value(pn544_dev -> iso_rst_gpio, 0);
                msleep(50);
                gpio_set_value(pn544_dev -> iso_rst_gpio, 1);
                msleep(50);
                pr_info("%s ISO RESET from SPI DONE\n", __func__);
#endif
            }
            else {
                pr_info("%s bad ese pwr arg %lu\n", __func__, arg);
                p61_access_unlock(pn544_dev);
                return - EBADRQC; /* Invalid request code */
            }
        }
        break;

        case P61_GET_PWR_STATUS:
        {
            p61_access_state_t current_state = P61_STATE_INVALID;
            p61_get_access_state(pn544_dev, &current_state);
            pr_info("%s: P61_GET_PWR_STATUS = %x", __func__, current_state);
            put_user(current_state, (int __user *)arg);
        }
        break;

        case PN544_SET_DWNLD_STATUS:
        {
            long ret;
            ret = set_jcop_download_state(arg);
            if(ret < 0)
            {
```

```c
                p61_access_unlock(pn544_dev);
                return ret;
            }
        }
    break;

    case P61_SET_WIRED_ACCESS:
    {
        p61_access_state_t current_state = P61_STATE_INVALID;
        p61_get_access_state(pn544_dev, &current_state);
        if (arg == 1)
        {
            if (current_state)
            {
                pr_info("%s : P61_SET_WIRED_ACCESS - enabling\n", __func__);
                p61_update_access_state(pn544_dev, P61_STATE_WIRED, true);
                if (current_state & P61_STATE_SPI_PRIO)
                {
                    if(pn544_dev->nfc_service_pid){
                        pr_info("nfc service pid %s    ---- %ld", __func__, pn544_dev->nfc_service_pid);
                        signal_handler(P61_STATE_SPI_PRIO, pn544_dev->nfc_service_pid);
                    }
                    else{
                        pr_info(" invalid nfc service pid....signalling failed %s   ---- %ld", __func__, pn544_dev->nfc_service_pid);
                    }
                }
                if((current_state & (P61_STATE_SPI|P61_STATE_SPI_PRIO)) == 0 && (pn544_dev->chip_pwr_scheme == PN67T_PWR_SCHEME))
                    gpio_set_value(pn544_dev->ese_pwr_gpio, 1);
            } else {
                pr_info("%s : P61_SET_WIRED_ACCESS - enabling failed \n", __func__);
                p61_access_unlock(pn544_dev);
                return -EBUSY; /* Device or resource busy */
            }
        } else if (arg == 0) {
```

```
                pr_info("%s : P61_SET_WIRED_ACCESS - disabling \n", __func__);
                if (current_state & P61_STATE_WIRED){
                    p61_update_access_state(pn544_dev, P61_STATE_WIRED, false);
                    if((current_state & (P61_STATE_SPI | P61_STATE_SPI_PRIO)) == 0 &&
(pn544_dev -> chip_pwr_scheme == PN67T_PWR_SCHEME))
                    {
                        svdd_sync_onoff(pn544_dev -> nfc_service_pid, P61_STATE_SPI_SVDD_SYNC_START);
                        gpio_set_value(pn544_dev -> ese_pwr_gpio, 0);
                        svdd_sync_onoff(pn544_dev -> nfc_service_pid, P61_STATE_SPI_SVDD_SYNC_END);
                    }
                } else {
                    pr_err("%s:P61_SET_WIRED_ACCESS - failed, current_state = %x \n",
                            __func__, pn544_dev -> p61_current_state);
                    p61_access_unlock(pn544_dev);
                    return -EPERM; /* Operation not permitted */
                }
            }
            else if(arg == 2)
            {
                pr_info("%s : P61_ESE_GPIO_LOW  \n", __func__);
                if(pn544_dev -> chip_pwr_scheme == PN67T_PWR_SCHEME)
                {
                    svdd_sync_onoff(pn544_dev -> nfc_service_pid, P61_STATE_SPI_SVDD_SYNC_START);
                    gpio_set_value(pn544_dev -> ese_pwr_gpio, 0);
                    svdd_sync_onoff(pn544_dev -> nfc_service_pid, P61_STATE_SPI_SVDD_SYNC_END);
                }
            }
            else if(arg == 3)
            {
                pr_info("%s : P61_ESE_GPIO_HIGH  \n", __func__);
                if(pn544_dev -> chip_pwr_scheme == PN67T_PWR_SCHEME)
                gpio_set_value(pn544_dev -> ese_pwr_gpio, 1);
            }
            else if(arg == 4)
```

```c
        {
            release_ese_lock(P61_STATE_WIRED);
        }
        else {
            pr_info("%s P61_SET_WIRED_ACCESS - bad arg %lu\n", __func__, arg);
            p61_access_unlock(pn544_dev);
            return -EBADRQC; /* Invalid request code */
        }
    }
    break;
case P544_SET_POWER_SCHEME:
    {
        if(arg == PN67T_PWR_SCHEME)
        {
            pn544_dev -> chip_pwr_scheme = PN67T_PWR_SCHEME;
            pr_info("%s : The power scheme is set to PN67T legacy \n", __func__);
        }
        else if(arg == PN80T_LEGACY_PWR_SCHEME)
        {
            pn544_dev -> chip_pwr_scheme = PN80T_LEGACY_PWR_SCHEME;
            pr_info("%s : The power scheme is set to PN80T_LEGACY_PWR_SCHEME,\n", __func__);
        }
        else if(arg == PN80T_EXT_PMU_SCHEME)
        {
            pn544_dev -> chip_pwr_scheme = PN80T_EXT_PMU_SCHEME;
            pr_info("%s : The power scheme is set to PN80T_EXT_PMU_SCHEME,\n", __func__);
        }
        else
        {
            pr_info("%s : The power scheme is invalid,\n", __func__);
        }
    }
    break;
case P544_SECURE_TIMER_SESSION:
    {
        secure_timer_operation(pn544_dev, arg);
```

```c
        }
        break;
    default:
        pr_err("%s bad ioctl %u\n", __func__, cmd);
        p61_access_unlock(pn544_dev);
        return -EINVAL;
    }
    p61_access_unlock(pn544_dev);
    pr_info("%s :exit cmd = %u, arg = %ld\n", __func__, cmd, arg);
    return 0;
}
EXPORT_SYMBOL(pn544_dev_ioctl);

static void secure_timer_workqueue(struct work_struct *Wq)
{
    p61_access_state_t current_state = P61_STATE_INVALID;
    printk(KERN_INFO "secure_timer_callback: called (%lu).\n", jiffies);
    /* Locking the critical section: ESE_PWR_OFF to allow eSE to shutdown peacefully :: START */
    get_ese_lock(P61_STATE_WIRED, MAX_ESE_ACCESS_TIME_OUT_MS);
    p61_update_access_state(pn544_dev, P61_STATE_SECURE_MODE, false);
    p61_get_access_state(pn544_dev, &current_state);

    if((current_state & (P61_STATE_SPI|P61_STATE_SPI_PRIO)) == 0)
    {
        printk(KERN_INFO "secure_timer_callback: make se_pwer_gpio low, state = %d", current_state);
        gpio_set_value(pn544_dev->ese_pwr_gpio, 0);
        /* Delay (2.5ms) after SVDD_PWR_OFF for the shutdown settlement time */
        usleep_range(2500, 3000);
        if(pn544_dev->nfc_service_pid == 0x00)
        {
            gpio_set_value(pn544_dev->ven_gpio, 0);
            printk(KERN_INFO "secure_timer_callback :make ven_gpio low, state = %d", current_state);
        }
    }
    pn544_dev->secure_timer_cnt = 0;
```

5 基于 Android 系统的 NFC 实现框架

```
    /* Locking the critical section: ESE_PWR_OFF to allow eSE to shutdown peacefully ::
END */
        release_ese_lock(P61_STATE_WIRED);
        return;
    }

    static void secure_timer_callback( unsigned long data )
    {
        /* Flush and push the timer callback event to the bottom half(work queue)
        to be executed later, at a safer time */
        flush_workqueue(pn544_dev -> pSecureTimerCbWq);
        queue_work(pn544_dev -> pSecureTimerCbWq, &pn544_dev -> wq_task);
        return;
    }

    static long start_seccure_timer(unsigned long timer_value)
    {
        long ret = -EINVAL;
        pr_info("start_seccure_timer: enter\n");
        /* Delete the timer if timer pending */
        if(timer_pending(&secure_timer) == 1)
        {
            pr_info("start_seccure_timer: delete pending timer \n");
            /* delete timer if already pending */
            del_timer(&secure_timer);
        }
        /* Start the timer if timer value is non-zero */
        if(timer_value)
        {
            init_timer(&secure_timer);
            setup_timer( &secure_timer, secure_timer_callback, 0 );

            pr_info(" start_seccure_timer: timeout % lums ( % lu)\n", timer_value,
jiffies);
            ret = mod_timer( &secure_timer, jiffies + msecs_to_jiffies(timer_value));
            if (ret)
                pr_info("start_seccure_timer: Error in mod_timer\n");
        }
```

```c
        return ret;
    }

    static long secure_timer_operation(struct pn544_dev * pn544_dev, unsigned long arg)
    {
        long ret = -EINVAL;
        unsigned long timer_value = arg;

        printk( KERN_INFO "secure_timer_operation, %d\n", pn544_dev -> chip_pwr_scheme);
        if(pn544_dev -> chip_pwr_scheme == PN80T_LEGACY_PWR_SCHEME)
        {
            ret = start_seccure_timer(timer_value);
            if(!ret)
            {
                pn544_dev -> secure_timer_cnt = 1;
                p61_update_access_state(pn544_dev, P61_STATE_SECURE_MODE, true);
            }
            else
            {
                pn544_dev -> secure_timer_cnt = 0;
                p61_update_access_state(pn544_dev, P61_STATE_SECURE_MODE, false);
                pr_info("%s :Secure timer reset \n", __func__);
            }
        }
        else
        {
            pr_info("%s :Secure timer session not applicable  \n", __func__);
        }
        return ret;
    }

    static long set_jcop_download_state(unsigned long arg)
    {
            p61_access_state_t current_state = P61_STATE_INVALID;
            long ret = 0;
            p61_get_access_state(pn544_dev, &current_state);
            pr_info("%s:Enter PN544_SET_DWNLD_STATUS:JCOP Dwnld state arg = %ld",__
```

```c
func__,arg);
            if(arg == JCP_DWNLD_INIT)
            {
                if(pn544_dev -> nfc_service_pid)
                {
                    pr_info("nfc service pid % s  ---- % ld",__func__,pn544_dev ->
nfc_service_pid);
                    signal_handler(JCP_DWNLD_INIT,pn544_dev -> nfc_service_pid);
                }
                else
                {
                    if (current_state & P61_STATE_JCP_DWNLD)
                    {
                        ret = - EINVAL;
                    }
                    else
                    {
                        p61_update_access_state(pn544_dev, P61_STATE_JCP_DWNLD, true);
                    }
                }
            }
            else if (arg == JCP_DWNLD_START)
            {
                if (current_state & P61_STATE_JCP_DWNLD)
                {
                    ret = - EINVAL;
                }
                else
                {
                    p61_update_access_state(pn544_dev, P61_STATE_JCP_DWNLD, true);
                }
            }
            else if (arg == JCP_SPI_DWNLD_COMPLETE)
            {
                if(pn544_dev -> nfc_service_pid)
                {
                    signal_handler(JCP_DWP_DWNLD_COMPLETE, pn544_dev -> nfc_service_pid);
                }
```

```c
                    p61_update_access_state(pn544_dev, P61_STATE_JCP_DWNLD, false);
            }
            else if (arg == JCP_DWP_DWNLD_COMPLETE)
            {
                    p61_update_access_state(pn544_dev, P61_STATE_JCP_DWNLD, false);
            }
            else
            {
                    pr_info("% s bad ese pwr arg % lu\n", __func__, arg);
                    p61_access_unlock(pn544_dev);
                    return - EBADRQC; /* Invalid request code */
            }
                    pr_info("% s: PN544_SET_DWNLD_STATUS = % x", __func__, current_state);

        return ret;
}


    int get_ese_lock(p61_access_state_t  p61_current_state, int timeout)
    {
        unsigned long tempJ = msecs_to_jiffies(timeout);
        if(down_timeout(&ese_access_sema, tempJ) != 0)
        {
            printk("get_ese_lock: timeout p61_current_state = % d\n", p61_current_state);
            return - EBUSY;
        }
        return 0;
    }
    EXPORT_SYMBOL(get_ese_lock);

    static void release_ese_lock(p61_access_state_t  p61_current_state)
    {
        up(&ese_access_sema);
    }



    static const struct file_operations pn544_dev_fops = {
            .owner = THIS_MODULE,
```

5 基于 Android 系统的 NFC 实现框架

```
        .llseek = no_llseek,
        .read = pn544_dev_read,
        .write = pn544_dev_write,
        .open = pn544_dev_open,
        .unlocked_ioctl = pn544_dev_ioctl,
};
#if DRAGON_NFC
static int pn544_parse_dt(struct device *dev,
    struct pn544_i2c_platform_data *data)
{
    struct device_node *np = dev->of_node;
    int errorno = 0;

#if !NEXUS5x
        data->irq_gpio = of_get_named_gpio(np, "nxp,pn544-irq", 0);
        if ((!gpio_is_valid(data->irq_gpio)))
            return -EINVAL;

        data->ven_gpio = of_get_named_gpio(np, "nxp,pn544-ven", 0);
        if ((!gpio_is_valid(data->ven_gpio)))
            return -EINVAL;

        data->firm_gpio = of_get_named_gpio(np, "nxp,pn544-fw-dwnld", 0);
        if ((!gpio_is_valid(data->firm_gpio)))
            return -EINVAL;

        data->ese_pwr_gpio = of_get_named_gpio(np, "nxp,pn544-ese-pwr", 0);
        if ((!gpio_is_valid(data->ese_pwr_gpio)))
            return -EINVAL;
        data->iso_rst_gpio = of_get_named_gpio(np, "nxp,pn544-iso-pwr-rst", 0);
        if ((!gpio_is_valid(data->iso_rst_gpio)))
            return -EINVAL;
#else
        data->ven_gpio = of_get_named_gpio_flags(np,
                                "nxp,gpio_ven", 0, NULL);
        data->firm_gpio = of_get_named_gpio_flags(np,
                                "nxp,gpio_mode", 0, NULL);
        data->irq_gpio = of_get_named_gpio_flags(np,
```

```
                                            "nxp,gpio_irq", 0, NULL);
    #endif
        pr_info("%s: %d, %d, %d, %d, %d error: %d\n", __func__,
            data -> irq_gpio, data -> ven_gpio, data -> firm_gpio, data -> iso_rst_gpio,
            data -> ese_pwr_gpio, errorno);


        return errorno;
    }
    #endif

    static int pn544_probe(struct i2c_client * client,
            const struct i2c_device_id * id)
    {
        int ret;
        struct pn544_i2c_platform_data * platform_data;
        //struct pn544_dev * pn544_dev;

    #if !DRAGON_NFC
        platform_data = client -> dev.platform_data;
    #else
        struct device_node * node = client -> dev.of_node;

        if (node) {
            platform_data = devm_kzalloc(&client -> dev,
                sizeof(struct pn544_i2c_platform_data), GFP_KERNEL);
            if (!platform_data) {
                dev_err(&client -> dev,
                    "nfc-nci probe: Failed to allocate memory\n");
                return -ENOMEM;
            }
            ret = pn544_parse_dt(&client -> dev, platform_data);
            if (ret)
            {
                pr_info("%s pn544_parse_dt failed", __func__);
            }
            client -> irq = gpio_to_irq(platform_data -> irq_gpio);
            if (client -> irq < 0)
            {
```

```c
            pr_info("%s gpio to irq failed", __func__);
        }
    } else {
        platform_data = client->dev.platform_data;
    }
#endif
    if (platform_data == NULL) {
        pr_err("%s : nfc probe fail\n", __func__);
        return -ENODEV;
    }

    if (!i2c_check_functionality(client->adapter, I2C_FUNC_I2C)) {
        pr_err("%s : need I2C_FUNC_I2C\n", __func__);
        return -ENODEV;
    }
#if !DRAGON_NFC
    ret = gpio_request(platform_data->irq_gpio, "nfc_int");
    if (ret)
        return -ENODEV;
    ret = gpio_request(platform_data->ven_gpio, "nfc_ven");
    if (ret)
        goto err_ven;
    ret = gpio_request(platform_data->ese_pwr_gpio, "nfc_ese_pwr");
    if (ret)
        goto err_ese_pwr;
    if (platform_data->firm_gpio) {
        ret = gpio_request(platform_data->firm_gpio, "nfc_firm");
        if (ret)
            goto err_firm;
    }
#ifdef ISO_RST
    if(platform_data->iso_rst_gpio) {
        ret = gpio_request(platform_data->iso_rst_gpio, "nfc_iso_rst");
        if (ret)
            goto err_iso_rst;
    }
#endif
#endif
```

```c
    pn544_dev = kzalloc(sizeof( * pn544_dev), GFP_KERNEL);
    if (pn544_dev = = NULL) {
        dev_err(&client -> dev,
                "failed to allocate memory for module data\n");
        ret = - ENOMEM;
        goto err_exit;
    }

    pn544_dev -> irq_gpio = platform_data -> irq_gpio;
    pn544_dev -> ven_gpio = platform_data -> ven_gpio;
    pn544_dev -> firm_gpio = platform_data -> firm_gpio;
    pn544_dev -> ese_pwr_gpio = platform_data -> ese_pwr_gpio;
#ifdef ISO_RST
    pn544_dev -> iso_rst_gpio = platform_data -> iso_rst_gpio;
#endif
    pn544_dev -> p61_current_state = P61_STATE_IDLE;
    pn544_dev -> nfc_ven_enabled = false;
    pn544_dev -> spi_ven_enabled = false;
    pn544_dev -> chip_pwr_scheme = PN67T_PWR_SCHEME;
    pn544_dev -> client = client;
    pn544_dev -> secure_timer_cnt = 0;

    ret = gpio_direction_input(pn544_dev -> irq_gpio);
    if (ret < 0) {
        pr_err(" % s :not able to set irq_gpio as input\n", __func__);
        goto err_ven;
    }
    ret = gpio_direction_output(pn544_dev -> ven_gpio, 0);
    if (ret < 0) {
        pr_err(" % s : not able to set ven_gpio as output\n", __func__);
        goto err_firm;
    }
    ret = gpio_direction_output(pn544_dev -> ese_pwr_gpio, 0);
    if (ret < 0) {
        pr_err(" % s : not able to set ese_pwr gpio as output\n", __func__);
        goto err_ese_pwr;
    }
    if (platform_data -> firm_gpio) {
```

```c
            ret = gpio_direction_output(pn544_dev -> firm_gpio, 0);
            if (ret < 0) {
                pr_err("%s : not able to set firm_gpio as output\n",
                    __func__);
                goto err_exit;
            }
        }
    #ifdef ISO_RST
        ret = gpio_direction_output(pn544_dev -> iso_rst_gpio, 0);
        if (ret < 0) {
            pr_err("%s : not able to set iso rst gpio as output\n", __func__);
            goto err_iso_rst;
        }
    #endif
        /* init mutex and queues */
        init_waitqueue_head(&pn544_dev -> read_wq);
        mutex_init(&pn544_dev -> read_mutex);
        sema_init(&ese_access_sema, 1);
        mutex_init(&pn544_dev -> p61_state_mutex);
        spin_lock_init(&pn544_dev -> irq_enabled_lock);
        pn544_dev -> pSecureTimerCbWq = create_workqueue(SECURE_TIMER_WORK_QUEUE);
        INIT_WORK(&pn544_dev -> wq_task, secure_timer_workqueue);
        pn544_dev -> pn544_device.minor = MISC_DYNAMIC_MINOR;
        pn544_dev -> pn544_device.name = "pn553";
        pn544_dev -> pn544_device.fops = &pn544_dev_fops;

        ret = misc_register(&pn544_dev -> pn544_device);
        if (ret) {
            pr_err("%s : misc_register failed\n", __FILE__);
            goto err_misc_register;
        }
        /* HiKey Compilation fix */
    #ifndef HiKey_620_COMPILATION_FIX
        wake_lock_init(&nfc_wake_lock, WAKE_LOCK_SUSPEND, "NFCWAKE");
    #endif
    #ifdef ISO_RST
        /* Setting ISO RESET pin high to power ESE during init */
        gpio_set_value(pn544_dev -> iso_rst_gpio, 1);
```

```c
#endif
    /* request irq. the irq is set whenever the chip has data available
     * for reading. it is cleared when all data has been read.
     */
    pr_info("%s : requesting IRQ %d\n", __func__, client->irq);
    pn544_dev->irq_enabled = true;
    ret = request_irq(client->irq, pn544_dev_irq_handler,
            IRQF_TRIGGER_HIGH, client->name, pn544_dev);
    if (ret) {
        dev_err(&client->dev, "request_irq failed\n");
        goto err_request_irq_failed;
    }
    enable_irq_wake(pn544_dev->client->irq);
    pn544_disable_irq(pn544_dev);
    i2c_set_clientdata(client, pn544_dev);
#if HWINFO
    /*
     * This function is used only if
     * hardware info is required during probe */
    check_hw_info();
#endif

    return 0;

err_request_irq_failed:
    misc_deregister(&pn544_dev->pn544_device);
err_misc_register:
    mutex_destroy(&pn544_dev->read_mutex);
    mutex_destroy(&pn544_dev->p61_state_mutex);
    kfree(pn544_dev);
err_exit:
    if (pn544_dev->firm_gpio)
        gpio_free(platform_data->firm_gpio);
err_firm:
    gpio_free(platform_data->ese_pwr_gpio);
err_ese_pwr:
    gpio_free(platform_data->ven_gpio);
err_ven:
```

```c
        gpio_free(platform_data->irq_gpio);
#ifdef ISO_RST
    err_iso_rst:
        gpio_free(platform_data->iso_rst_gpio);
#endif
        return ret;
}

static int pn544_remove(struct i2c_client *client)
{
        struct pn544_dev *pn544_dev;

        pn544_dev = i2c_get_clientdata(client);
        free_irq(client->irq, pn544_dev);
        misc_deregister(&pn544_dev->pn544_device);
        mutex_destroy(&pn544_dev->read_mutex);
        mutex_destroy(&pn544_dev->p61_state_mutex);
        gpio_free(pn544_dev->irq_gpio);
        gpio_free(pn544_dev->ven_gpio);
        gpio_free(pn544_dev->ese_pwr_gpio);
        destroy_workqueue(pn544_dev->pSecureTimerCbWq);
#ifdef ISO_RST
        gpio_free(pn544_dev->iso_rst_gpio);
#endif
        pn544_dev->p61_current_state = P61_STATE_INVALID;
        pn544_dev->nfc_ven_enabled = false;
        pn544_dev->spi_ven_enabled = false;

        if (pn544_dev->firm_gpio)
            gpio_free(pn544_dev->firm_gpio);
        kfree(pn544_dev);

        return 0;
}

static const struct i2c_device_id pn544_id[] = {
#if NEXUS5x
        { "pn548", 0 },
```

```c
#else
        { "pn544", 0 },
#endif
        { }
};
#if DRAGON_NFC
static struct of_device_id pn544_i2c_dt_match[] = {
    {
#if NEXUS5x
        .compatible = "nxp,pn548",
#else
        .compatible = "nxp,pn544",
#endif
    },
    {}
};
#endif
static struct i2c_driver pn544_driver = {
        .id_table = pn544_id,
        .probe = pn544_probe,
        .remove = pn544_remove,
        .driver = {
                .owner = THIS_MODULE,
#if NEXUS5x
                .name = "pn548",
#else
                .name = "pn544",
#endif
#if DRAGON_NFC
                .of_match_table = pn544_i2c_dt_match,
#endif
        },
};
#if HWINFO
/*******************************************************
 * Function         check_hw_info
 *
 * Description      This function is called during pn544_probe to retrieve
```

```
 *                  HW info.
 *                  Useful get HW information in case of previous FW download is
 *                  interrupted and core reset is not allowed.
 *                  This function checks if core reset is allowed, if not
 *                  sets DWNLD_REQ(firm_gpio), ven reset and sends firmware
 *                  get version command.
 *                  In response HW information will be received.
 *
 * Returns          None
 *
 ********************************************************************/
static void check_hw_info() {
    char read_data[20];
    int ret, get_version_len = 8, retry_count = 0;
    static uint8_t cmd_reset_nci[] = {0x20, 0x00, 0x01, 0x00};
    char get_version_cmd[] =
    {0x00, 0x04, 0xF1, 0x00, 0x00, 0x00, 0x6E, 0xEF};

    pr_info("%s :Enter\n", __func__);

    /*
     * Ven Reset   before sending core Reset
     * This is to check core reset is allowed or not.
     * If not allowed then previous FW download is interrupted in between
     * */
    pr_info("%s :Ven Reset \n", __func__);
    gpio_set_value(pn544_dev -> ven_gpio, 1);
    msleep(10);
    gpio_set_value(pn544_dev -> ven_gpio, 0);
    msleep(10);
    gpio_set_value(pn544_dev -> ven_gpio, 1);
    msleep(10);
    ret = i2c_master_send(pn544_dev -> client, cmd_reset_nci, 4);

    if (ret == 4) {
        pr_info("%s : core reset write success\n", __func__);
    } else {
```

```c
/*
 * Core reset  failed.
 * set the DWNLD_REQ , do ven reset
 * send firmware download info command
 * */
pr_err("%s : write failed\n", __func__);
pr_info("%s power on with firmware\n", __func__);
gpio_set_value(pn544_dev -> ven_gpio, 1);
msleep(10);
if (pn544_dev -> firm_gpio) {
    p61_update_access_state(pn544_dev, P61_STATE_DWNLD, true);
    gpio_set_value(pn544_dev -> firm_gpio, 1);
}
msleep(10);
gpio_set_value(pn544_dev -> ven_gpio, 0);
msleep(10);
gpio_set_value(pn544_dev -> ven_gpio, 1);
msleep(10);
ret = i2c_master_send(pn544_dev -> client, get_version_cmd, get_version_len);
if (ret != get_version_len) {
    ret = -EIO;
    pr_err("%s : write_failed \n", __func__);
}
else {
    pr_info("%s :data sent\n", __func__);
}

ret = 0;

while (retry_count < 10) {

    /*
     * Wait for read interrupt
     * If spurious interrupt is received retry again
     * */
    pn544_dev-> irq_enabled = true;
    enable_irq(pn544_dev -> client -> irq);
    enable_irq_wake(pn544_dev -> client -> irq);
```

```
            ret = wait_event_interruptible(
                pn544_dev - > read_wq,
                !pn544_dev - > irq_enabled);

            pn544_disable_irq(pn544_dev);

            if (gpio_get_value(pn544_dev - > irq_gpio))
                break;

            pr_warning(" % s: spurious interrupt detected\n", __func__);
            retry_count ++ ;
        }

        if(ret) {
            return;
        }

        /*
         * Read response data and copy into hw_type_info
         * */
        ret = i2c_master_recv(pn544_dev - > client, read_data, 14);

        if(ret) {
            memcpy(hw_info.data, read_data, ret);
            hw_info.len = ret;
            pr_info(" % s :data received len    : % d\n", __func__,hw_info.len);
        }
        else {
            pr_err(" % s :Read Failed\n", __func__);
        }
    }
}
# endif
/*
 * module load/unload record keeping
 */

static int __init pn544_dev_init(void)
```

```c
{
    pr_info("Loading pn544 driver\n");
    return i2c_add_driver(&pn544_driver);
}
module_init(pn544_dev_init);

static void __exit pn544_dev_exit(void)
{
    pr_info("Unloading pn544 driver\n");
    i2c_del_driver(&pn544_driver);
}
module_exit(pn544_dev_exit);

MODULE_AUTHOR("Sylvain Fonteneau");
MODULE_DESCRIPTION("NFC PN544 driver");
MODULE_LICENSE("GPL");
```

P61通过SPI总线连接到REE的驱动代码示例如下。

头文件p61.h如下：

```c
/*
 * Copyright (C) 2015 NXP Semiconductors
 *
 * Licensed under the Apache License, Version 2.0 (the "License");
 * you may not use this file except in compliance with the License.
 * You may obtain a copy of the License at
 *
 *     http://www.apache.org/licenses/LICENSE-2.0
 *
 * Unless required by applicable law or agreed to in writing, software
 * distributed under the License is distributed on an "AS IS" BASIS,
 * WITHOUT WARRANTIES OR CONDITIONS OF ANY KIND, either express or implied.
 * See the License for the specific language governing permissions and
 * limitations under the License.
 */

#define P61_MAGIC 0xEA
#define P61_SET_PWR _IOW(P61_MAGIC, 0x01, unsigned int)
#define P61_SET_DBG _IOW(P61_MAGIC, 0x02, unsigned int)
```

#define P61_SET_POLL _IOW(P61_MAGIC, 0x03, unsigned int)

```
struct p61_spi_platform_data {
    unsigned int irq_gpio;
    unsigned int rst_gpio;
};
```

源文件 p61.c 如下：

```
/*
 * Copyright (C) 2010 Trusted Logic S.A.
 *
 * This program is free software; you can redistribute it and/or modify
 * it under the terms of the GNU General Public License as published by
 * the Free Software Foundation; either version 2 of the License, or
 * (at your option) any later version.
 *
 * This program is distributed in the hope that it will be useful,
 * but WITHOUT ANY WARRANTY; without even the implied warranty of
 * MERCHANTABILITY or FITNESS FOR A PARTICULAR PURPOSE.  See the
 * GNU General Public License for more details.
 *
 * You should have received a copy of the GNU General Public License
 * along with this program; if not, write to the Free Software
 * Foundation, Inc., 51 Franklin St, Fifth Floor, Boston, MA 02110-1301 USA
 *
 */

#include <linux/kernel.h>
#include <linux/module.h>
#include <linux/fs.h>
#include <linux/slab.h>
#include <linux/init.h>
#include <linux/list.h>
#include <linux/irq.h>
#include <linux/jiffies.h>
#include <linux/uaccess.h>
#include <linux/delay.h>
#include <linux/interrupt.h>
```

```
#include < linux/io.h >
#include < linux/platform_device.h >
#include < linux/gpio.h >
#include < linux/miscdevice.h >
#include < linux/spinlock.h >
#include < linux/spi/spi.h >
#include < linux/sched.h >
#include < linux/poll.h >
#include < linux/timer.h >

#define P61_DBG_LEVEL 0
//#define TIMER_ENABLE
#define TIMER_ENABLE 0
#define IRQ_ENABLE 1
#if P61_DBG_LEVEL
#define NFC_DBG_MSG(msg...) printk(KERN_ERR "[NFC PN61] : " msg);
#else
#define NFC_DBG_MSG(msg...)
#endif

#define NFC_ERR_MSG(msg...) printk(KERN_ERR "[NFC PN61] : " msg);

#define P61_IRQ      24
#define P61_RST      62
#define P61_PWR      16
#define P61_CS       90

#define P61_mosi     0
#define P61_somi     1
#define P61_clk      3

#define NFC_p61_rst    GPIO_CFG(P61_RST, 0, GPIO_CFG_OUTPUT, GPIO_CFG_NO_PULL, GPIO_CFG_2MA)

#define P61_MAGIC 0xE8
/*
 * PN544 power control via ioctl
 * PN544_SET_PWR(0): power off
```

* PN544_SET_PWR(1): power on

* PN544_SET_PWR(2): reset and power on with firmware download enabled */

```c
#define P61_SET_PWR _IOW(P61_MAGIC, 0x01, unsigned int)
#define MAX_BUFFER_SIZE 4096 //4k
int sendFrame(struct file * filp, const char __user data[], char mode, int count);
int sendChainedFrame(struct file * filp, const unsigned char __user data[], int count);
static int p61_dev_close(void);
unsigned char * apduBuffer;
unsigned char * apduBuffer1;
unsigned char * gRecvBuff;
unsigned char * checksum;
int apduBufferidx = 0;
int apduBufferlen = 256;
const char PH_SCAL_T1_CHAINING = 0x20;
const char PH_SCAL_T1_SINGLE_FRAME = 0x00;
const char PH_SCAL_T1_R_BLOCK = 0x80;
const char PH_SCAL_T1_S_BLOCK = 0xC0;
const char PH_SCAL_T1_HEADER_SIZE_NO_NAD = 0x02;
static unsigned char seqCounterCard = 0;
static unsigned char seqCounterTerm = 1;
short ifs = 254;
short headerSize = 3;
unsigned char sof = 0xA5;
unsigned char csSize = 1;
static unsigned char array[];
const char C_TRANSMIT_NO_STOP_CONDITION = 0x01;
const char C_TRANSMIT_NO_START_CONDITION = 0x02;
const char C_TRANSMIT_NORMAL_SPI_OPERATION = 0x04;

typedef struct respData {
    unsigned char * data;
    int len;
}respData_t;

respData_t * gStRecvData = NULL;
unsigned char * gSendframe = NULL;
unsigned char * gDataPackage = NULL;
unsigned char * data1 = NULL;
```

```c
#define MEM_CHUNK_SIZE (256)
unsigned char * lastFrame = NULL;
int lastFrameLen;
void init(void);
void setAddress(short address);
void setBitrate(short bitrate);
int nativeSetAddress(short address);
int nativeSetBitrate(short bitrate);
unsigned char helperComputeLRC(unsigned char data[], int offset, int length);
void receiveAcknowledge(struct file * filp);
void receiveAndCheckChecksum(struct file * filp,short rPcb, short rLen, unsigned char data[],int len);
respData_t * receiveHeader(struct file * filp);
respData_t * receiveFrame(struct file * filp,short rPcb, short rLen);
int send(struct file * filp,unsigned char * * data,unsigned char mode,int len);
int receive(struct file * filp,unsigned char * * data,int len, unsigned char mode);
respData_t * receiveChainedFrame(struct file * filp,short rPcb, short rLen);
void sendAcknowledge(struct file * filp);
static ssize_t p61_dev_write(struct file * filp, const char __user * buf,
            size_t count, loff_t * offset);
static ssize_t p61_dev_read(struct file * filp, char __user * buf,
            size_t count, loff_t * offset);
static ssize_t p61_dev_receiveData(struct file * filp, char __user * buf,
            size_t count, loff_t * offset);
/* static ssize_t p61_dev_sendData(struct file * filp, const char __user * buf,
            size_t count, loff_t * offset);
 */
static respData_t * p61_dev_receiveData_internal(struct file * filp);


poll_table * wait;

struct p61_dev{
    wait_queue_head_t    read_wq;
    struct mutex    read_mutex;
    struct spi_device    * spi;
    struct miscdevice    p61_device;
    char * rp, * wp;
    unsigned int    irq_gpio; // IRQ will be used later (P61 will interrupt DH for any ntf
```

```c
        bool      irq_enabled;
        spinlock_t      irq_enabled_lock;
        wait_queue_head_t * inq, * outq;
        int buffersize;
        struct spi_message msg;
        struct spi_transfer transfer;
        bool firm_gpio;
        bool ven_gpio;
};

struct p61_control {
        struct spi_message msg;
        struct spi_transfer transfer;
        unsigned char * tx_buff;
        unsigned char * rx_buff;
};


static int timer_expired = 0;
#ifdef TIMER_ENABLE
static struct timer_list recovery_timer;
static int timer_started = 0;
static void p61_disable_irq(struct p61_dev * p61_dev);
void my_timer_callback( unsigned long data )
{
        timer_expired = 1;
}

static int start_timer(void)
{
        int ret;
        setup_timer( &recovery_timer, my_timer_callback, 0 );
        ret = mod_timer( &recovery_timer, jiffies + msecs_to_jiffies(2000) );
        if (ret) {
                printk(KERN_INFO "Error in mod_timer\n");
        } else {
                timer_started = 1;
        }
        return 0;
```

```c
    }

    void cleanup_timer( void )
    {
        int ret;

        ret = del_timer( &recovery_timer );
        return;
    }
    #endif

    static int p61_dev_open(struct inode * inode, struct file * filp)
    {
        struct p61_dev * p61_dev = container_of(filp - > private_data,
                struct p61_dev,
                p61_device);
        filp - > private_data = p61_dev;
        gRecvBuff = (unsigned char * )kmalloc(300,GFP_KERNEL);
        gStRecvData = (respData_t * ) kmalloc(sizeof(respData_t),GFP_KERNEL);
        checksum = (unsigned char * )kmalloc(csSize,GFP_KERNEL);
        gSendframe = (unsigned char * )kmalloc(300,GFP_KERNEL);
        gDataPackage = (unsigned char * )kmalloc(300,GFP_KERNEL);
        apduBuffer = (unsigned char * ) kmalloc(apduBufferlen,GFP_KERNEL);
        data1 = (unsigned char * )kmalloc(1,GFP_KERNEL);
        NFC_DBG_MSG("% s : Major No: % d, Minor No: % d\n", __func__, imajor(inode), iminor(inode));
        return 0;
    }

    static int p61_dev_close(void)
    {
        if(checksum ! = NULL)
        kfree(checksum);
        if(gRecvBuff ! = NULL)
        kfree(gRecvBuff);
        if(gStRecvData ! = NULL)
        kfree(gStRecvData);
        if(gSendframe ! = NULL)
```

```
        kfree(gSendframe);
        if(gDataPackage != NULL)
        kfree(gDataPackage);
        if(apduBuffer != NULL)
        kfree(apduBuffer);
        if(data1 != NULL)
        kfree(data1);
        return 0;
    }

    void SpiReadSingle(struct spi_device * spi, unsigned char * pTxbuf, unsigned char *
pbuf, unsigned int length)
    {
        while (length) {
            length--;
            spi_write_then_read(spi, pTxbuf, 1, pbuf, 1);
            pbuf++;
            pTxbuf++;
        }
    }
    //static unsigned int p61_dev_poll(struct file * filp, char * buf,
    //        size_t count, loff_t * offset)

    static unsigned int p61_dev_poll (struct file * filp, struct poll_table_struct * buf)
    {
        int mask = -1;
        int left = 0;
        struct p61_dev * p61_dev = filp->private_data;
        NFC_DBG_MSG(KERN_INFO "p61_dev_poll called \n");
        left = (p61_dev->rp + p61_dev->buffersize - p61_dev->wp) % (p61_dev->
buffersize);
        poll_wait(filp,p61_dev->inq,wait);
        if(p61_dev->rp != p61_dev->wp)
            mask|= POLLIN|POLLRDNORM;
        if(left!= 1)
            mask|= POLLOUT|POLLRDNORM;
        return mask;
    }
```

```c
/**
 * Entry point function for receiving data. Based on the PCB byte this function
 * either receives a single frame or a chained frame.
 *
 */
/* static ssize_t p61_dev_receiveData(struct file * filp, char * buf,
        size_t count, loff_t * offset)
        */
static ssize_t p61_dev_receiveData(struct file * filp, char __user * buf,
        size_t count, loff_t * offset)
{
    respData_t * rsp = p61_dev_receiveData_internal(filp);
    //if(count > = rsp - > len)
    //{
        count = rsp - > len;
    //}
    if (0 < count) {
        //memcpy(buf,rsp - > data,count);
        if (copy_to_user(buf, rsp - > data, count)) {
            NFC_DBG_MSG(" % s : failed to copy to user space\n", __func__);
            return - EFAULT;
        } else {
            //Nothing
        }
    }

    return count;
}

static respData_t * p61_dev_receiveData_internal(struct file * filp)
{
    short rPcb = 0;
    short rLen = 0;
    respData_t * header = NULL;
    respData_t * respData = NULL;
    unsigned char * wtx = NULL;
    unsigned char * data = NULL;
    //unsigned char * data1 = NULL;
```

5 基于 Android 系统的 NFC 实现框架

```c
            int len = 0;
            int len1 = 0;
            int stat_timer;
    Start:
            NFC_DBG_MSG(KERN_INFO   "receiveData - Enter\n");

            // receive the T = 1 header
            header = (respData_t * )receiveHeader(filp);
            if(header == NULL)
            {
                NFC_ERR_MSG(KERN_ALERT "ERROR:Failed to receive header data\n");
                return NULL;
            }
            rPcb = header -> data[0];
            rLen = (short) (header -> data[1] & 0xFF);
            NFC_DBG_MSG(KERN_ALERT "receive header data rPcb = 0x%x , rLen = %d\n",rPcb,
    rLen);

        #if 1

            //check the header if wtx is requested
            if ((rPcb & PH_SCAL_T1_S_BLOCK) == PH_SCAL_T1_S_BLOCK)
            {
                NFC_DBG_MSG(KERN_ALERT "receiveDatav - WTX requested\n");
                data = gRecvBuff;
                len = 1;
                NFC_DBG_MSG(KERN_ALERT "receiveDatav - WTX1 requested\n");
                receive(filp,&data,len, C_TRANSMIT_NO_STOP_CONDITION | C_TRANSMIT_NO_START_
    CONDITION);
                NFC_DBG_MSG(KERN_ALERT "receiveDatav - WTX2 requested\n");
                receiveAndCheckChecksum(filp,rPcb, rLen, data,len);
                NFC_DBG_MSG(KERN_ALERT "receiveDatav - WTX3 requested\n");
                NFC_DBG_MSG(KERN_ALERT "value is %x %x",data[0],data[1]);
                memset(gRecvBuff,0,300);
                wtx = gRecvBuff;
                wtx[0] = 0x00; wtx[1] = 0xE3; wtx[2] = 0x01; wtx[3] = 0x01; wtx[4] = 0xE3;
                len1 = 5;
                udelay(1000);
```

```c
            send(filp,&wtx, C_TRANSMIT_NORMAL_SPI_OPERATION,len1);
            udelay(1000);

            //gStRecvData -> len = 5;
            //memcpy(gStRecvData -> data, wtx, 5);;
#ifdef TIMER_ENABLE
            stat_timer = start_timer();
#endif
            //return gStRecvData;
            goto Start;
        }


        //check the header if retransmit is requested
        if ((rPcb & PH_SCAL_T1_R_BLOCK) == PH_SCAL_T1_R_BLOCK)
        {
            memset(data1,0,1);
            len1 = 1;
            receiveAndCheckChecksum(filp,rPcb, rLen, data1,len1);
            udelay(1000);
            send(filp,&lastFrame, C_TRANSMIT_NORMAL_SPI_OPERATION,lastFrameLen);
            udelay(1000);
            goto Start;
            //return (ssize_t)p61_dev_receiveData_internal(filp);
        }


        //check the PCB byte and receive the rest of the frame
        if ((rPcb & PH_SCAL_T1_CHAINING) == PH_SCAL_T1_CHAINING)
        {
            NFC_DBG_MSG(KERN_ALERT "Chained Frame Requested\n");
            return receiveChainedFrame(filp,rPcb, rLen);
        }
        else
        {
            NFC_DBG_MSG(KERN_ALERT "receiveFrame Requested\n");
            respData = receiveFrame(filp,rPcb, rLen);
            NFC_DBG_MSG(KERN_ALERT " ************* 0x%x \n",respData -> data[0]);
            return respData;
        }
```

```
#endif
    return NULL;
}
/**
 * This function is used to receive a single T = 1 frame
 *
 * @param rPcb
 *          PCB field of the current frame
 * @param rLen
 *          LEN field of the current frame
 * @param filp
 *          File pointer
 */

respData_t * receiveFrame(struct file * filp,short rPcb, short rLen)
{

    int status = 0;
    respData_t * respData = NULL;
    NFC_DBG_MSG(KERN_ALERT "receiveFrame - Enter\n");
    respData = gStRecvData;
    respData -> data = gRecvBuff;
    respData -> len = rLen;
    // modify the card send sequence counter
    seqCounterCard = (seqCounterCard ^ 1);

    // receive the DATA field and check the checksum
    status = receive(filp,&(respData -> data),respData -> len, C_TRANSMIT_NO_STOP_CONDITION | C_TRANSMIT_NO_START_CONDITION);

    receiveAndCheckChecksum(filp,rPcb, rLen, respData -> data,respData -> len);

    NFC_DBG_MSG(KERN_ALERT "receiveFrame - Exit\n");

    return respData;
}

/**
```

```
 *  This function is used to receive a chained frame.
 *
 *  @param rPcb
 *          PCB field of the current frame
 *  @param rLen
 *          LEN field of the current frame
 *  @param filp
 *          File pointer
 */

respData_t * receiveChainedFrame(struct file * filp,short rPcb, short rLen)
{
    respData_t * data_rec = NULL;
    respData_t * header = NULL ;
    respData_t * respData = NULL;
    respData_t * apdbuff = NULL;
    NFC_DBG_MSG(KERN_ALERT "receiveChainedFrame - Enter\n");
    // receive a chained frame as long as chaining is indicated in the PCB
    do {
        // receive the DATA field of the current frame
        NFC_DBG_MSG(KERN_ALERT "p61_dev_read - test4 count [0x%x] \n",rLen);
        data_rec = receiveFrame(filp,rPcb, rLen);
        // write it into an apduBuffer memory
        memcpy((apduBuffer + apduBufferidx),data_rec -> data,data_rec -> len);

        //update the index to next free slot
        apduBufferidx += data_rec -> len;

        // send the acknowledge for the current frame
        udelay(1000);
        sendAcknowledge(filp);
        udelay(1000);
        // receive the header of the next frame
        header = receiveHeader(filp);

        rPcb = header -> data[0];
        rLen = (header -> data[1] & 0xFF);
```

```c
    }while((rPcb & PH_SCAL_T1_CHAINING) == PH_SCAL_T1_CHAINING);

    // receive the DATA field of the last frame
    respData = receiveFrame(filp,rPcb, rLen);
    memcpy(apduBuffer + apduBufferidx,respData -> data,respData -> len);
    //update the index to next free slot
    apduBufferidx += respData -> len;

    // return the entire received apdu
    apdbuff = (respData_t * )kmalloc(sizeof(respData_t),GFP_KERNEL);
    if(apdbuff == NULL) {
        NFC_ERR_MSG(KERN_ALERT "receiveChainedFrame 2 - KMALLOC FAILED!!!\n");
        return NULL;
    }

    apdbuff -> data = (unsigned char * )kmalloc(apduBufferidx,GFP_KERNEL);
    if(apdbuff -> data == NULL) {
        NFC_ERR_MSG(KERN_ALERT "receiveChainedFrame 3 - KMALLOC FAILED!!!\n");
        return NULL;
    }
    memcpy(apdbuff -> data,apduBuffer,apduBufferidx);
    apdbuff -> len = apduBufferidx;

    NFC_DBG_MSG(KERN_ALERT "receiveChainedFrame - Exit\n");
    return apdbuff;
}
/ * *
 * This function is used to send an acknowledge for an received I frame
 * in chaining mode.
 *
 * /
void sendAcknowledge(struct file * filp)
{
    unsigned char * ack = NULL;
    NFC_DBG_MSG(KERN_ALERT "sendAcknowledge - Enter\n");
    ack = gSendframe;
    // prepare the acknowledge and send it
    NFC_DBG_MSG(KERN_ALERT "seqCounterCard value is [0x % x]\n",seqCounterCard);
```

```
        ack[0] = 0x00;
        ack[1] = (unsigned char)(PH_SCAL_T1_R_BLOCK | (unsigned char)(seqCounterCard << 4));
        ack[2] = 0x00;
        ack[3] = helperComputeLRC(ack, 0, (sizeof(ack) / sizeof(ack[0])) - 2);

        send(filp,&ack, C_TRANSMIT_NORMAL_SPI_OPERATION,sizeof(ack)/sizeof(ack[0]));

        NFC_DBG_MSG(KERN_ALERT "sendAcknowledge - Exit\n");

}
/**
  * This function sends either a chained frame or a single T = 1 frame
  *
  * @param buf
  *              the data to be send
  *
  */
/* static ssize_t p61_dev_sendData(struct file * filp, unsigned char * buf,
        size_t count, loff_t * offset)
        */
static ssize_t p61_dev_sendData(struct file * filp, const char __user * buf,
        size_t count, loff_t * offset)
{
    int ret = -1;
    int i;
    init();
    printk("p61_dev_sendData %d - Enter \n", (int)count);
    if(count < = ifs)
    {
        for(i = 0;i < = count;i ++ ) {
            NFC_ERR_MSG("p61_dev_sendData : buf%d[0x%x]",i,buf[i]);
        }
        ret = sendFrame(filp,buf, PH_SCAL_T1_SINGLE_FRAME,count);
        printk("Vaue of count_status is %d \n",ret);
    }
    else
    {
        //return sendChainedFrame(data);
```

```c
        ret = sendChainedFrame(filp,buf,count);
    }
    NFC_DBG_MSG(KERN_INFO "p61_dev_sendData: count_status is %d \n",ret);
    return ret;
}

static long  p61_dev_ioctl(struct file * filp, unsigned int cmd,
            unsigned long arg)
{
    int ret;
    struct p61_dev * p61_dev = NULL;
    uint8_t buf[100];
    //NFC_DBG_MSG(KERN_ALERT "p61_dev_ioctl - Enter %x arg = 0x%x\n",cmd, arg);
    p61_dev = filp -> private_data;
    p61_dev -> ven_gpio = P61_RST;
    printk("p61_dev_ioctl enter\n");

    switch (cmd) {
    case P61_SET_PWR:
        NFC_DBG_MSG(KERN_ALERT "P61_SET_PWR - Enter P61_RST = 0x%x\n", P61_RST);
        if (arg == 2) {
            printk("p61_dev_ioctl    download firmware \n");
            /* power on with firmware download (requires hw reset) */
            gpio_direction_output(P61_RST, 1);
            NFC_DBG_MSG(KERN_ALERT "p61_dev_ioctl - 1\n");
            msleep(20);
            gpio_direction_output(P61_RST, 0);
            NFC_DBG_MSG(KERN_ALERT "p61_dev_ioctl - 0\n");
            msleep(50);
            ret = spi_read (p61_dev -> spi,(void * ) buf, sizeof(buf));
            msleep(50);
            gpio_direction_output(P61_RST, 1);
            NFC_DBG_MSG(KERN_ALERT "p61_dev_ioctl - 1 \n");
            msleep(20);

        } else if (arg == 1) {
            printk("p61_dev_ioctl    power on \n");
            /* power on */
```

```c
                NFC_DBG_MSG(KERN_ALERT "p61_dev_ioctl - 1 (arg = 1)\n");
                gpio_direction_output(P61_RST, 1);

        } else if (arg == 0) {
                printk("p61_dev_ioctl    power off \n");
                /* power off */
                NFC_DBG_MSG(KERN_ALERT "p61_dev_ioctl - 0 (arg = 0)\n");
                gpio_direction_output(P61_RST, 0);
                udelay(100);
        } else {
                return - EINVAL;
        }
        break;
    default:
        return - EINVAL;
    }

    return 0;
}


/**
 * This function is used to send a chained frame.
 *
 * @param data
 *          the data to be send
 */
int sendChainedFrame(struct file * filp,const unsigned char data[],int len)
{
    int count_status = 0 ;
    int length = len;
    int offset = 0;
    int ret = 0;
    unsigned char * lastDataPackage = NULL;
    unsigned char * dataPackage = NULL;
    NFC_DBG_MSG(KERN_INFO "sendChainedFrame - Enter\n");
    dataPackage = gDataPackage;
    do
    {
```

5 基于 Android 系统的 NFC 实现框架

```c
        NFC_DBG_MSG(KERN_INFO "sendChainedFrame \n");
        // send a chained frame and receive the acknowledge
        memcpy(&dataPackage[0],&data[offset], ifs);

        count_status = sendFrame(filp,dataPackage, PH_SCAL_T1_CHAINING,ifs);
        if(count_status == 0)
        {
            NFC_ERR_MSG(KERN_INFO "ERROR1: Failed to send Frame\n");
            return -1;
        }
        receiveAcknowledge(filp);
        udelay(1000);
        length = length - ifs;
        offset = offset + ifs;
        ret += count_status;
    }while (length > ifs);

#if 1
    // send the last frame
    lastDataPackage = gDataPackage;
    memcpy(&lastDataPackage[0],&data[offset], length);

    count_status = sendFrame(filp,lastDataPackage, PH_SCAL_T1_SINGLE_FRAME,length);

    if(count_status == 0) {
        NFC_ERR_MSG(KERN_INFO "ERROR2:Failed to send Frame\n");
        return -1;
    }
#endif
    NFC_DBG_MSG(KERN_INFO "sendChainedFrame - Exit\n");
    ret += count_status;
    return ret;
}
/**
 * This function is used to receive an Acknowledge of an I frame
 *
 */
void receiveAcknowledge(struct file *filp)
```

```c
{
    respData_t * header = NULL;
    short rPcb = 0;
    short rLen = 0;
    int len = 1;
    unsigned char * cs = NULL;
    NFC_DBG_MSG(KERN_INFO "receiveAcknowledge - Enter\n");
    cs = gRecvBuff;
    header = (respData_t *) receiveHeader(filp);
    rPcb = (header -> data[0] & 0xFF);
    rLen = (header -> data[1] & 0xFF);
    receiveAndCheckChecksum(filp,rPcb, rLen, cs,len);
    NFC_DBG_MSG(KERN_ALERT "receiveAcknowledge - Exit\n");
}


/**
 * This function is used to receive the header of the next T = 1 frame.
 * If no data is available the function polls the data line as long as it receives the
 * start of the header and then receives the entire header.
 *
 */
respData_t * receiveHeader(struct file * filp)
{
    int count_status = 0;
    //unsigned char * ready = NULL;
    respData_t * header = NULL;
    unsigned char * r_frame = NULL;
    int len = 1;
    NFC_DBG_MSG(KERN_ALERT "receiveHeader - Enter\n");
    header = gStRecvData;
    header -> data = gRecvBuff;
    header -> len = PH_SCAL_T1_HEADER_SIZE_NO_NAD;
    count_status = receive(filp,&gRecvBuff,len, C_TRANSMIT_NO_STOP_CONDITION);
    NFC_DBG_MSG(KERN_ALERT "sof is :0x%x\n",gRecvBuff[0]);
    // check if we received ready
#ifdef TIMER_ENABLE
again:
#endif
```

```
            while(gRecvBuff[0] != sof)
            {
                NFC_DBG_MSG(KERN_ALERT "SOF not found\n");
                // receive one byte and keep SS line low
                count_status = receive(filp,&gRecvBuff,len, C_TRANSMIT_NO_STOP_CONDITION |
C_TRANSMIT_NO_START_CONDITION);
                NFC_DBG_MSG(KERN_ALERT "in While SOF is : 0x%x \n",gRecvBuff[0]);
            }
        #ifdef TIMER_ENABLE
            if (timer_started) {
                timer_started = 0;
                cleanup_timer();
            }
            if (timer_expired == 1) {
                memset(gSendframe,0,300);
                r_frame = gSendframe;
                r_frame[0] = 0x00;r_frame[1] = 0x00;r_frame[2] = 0x00;r_frame[3] = 0x00;
                timer_started = 0;
                timer_expired = 0;
                cleanup_timer();
                // printk(KERN_INFO "************ Sending R Frame \n");
                send(filp,&r_frame, C_TRANSMIT_NORMAL_SPI_OPERATION, 4);
                goto again;
            }
        #endif
            NFC_DBG_MSG(KERN_ALERT "SOF FOUND\n");
            // we received ready byte, so we can receive the rest of the header and keep SS
line low
            count_status = receive(filp,&(header -> data),header -> len , C_TRANSMIT_NO_STOP_
CONDITION | C_TRANSMIT_NO_START_CONDITION);
            NFC_DBG_MSG(KERN_ALERT "receiveHeader - Exit\n");

            return header;
        }
        /**
        * This function is used to receive and check the checksum of the T = 1 frame.
        *
        * @param rPcb
```

```c
 *                  PCB field of the current frame
 * @param rLen
 *                  LEN field of the current frame
 * @param data
 *                  DATA field of the current frame
 * @param dataLength
 *
 * @param filp
 *                  File pointer
 *
 */
void receiveAndCheckChecksum(struct file * filp,short rPcb, short rLen, unsigned char data[],int dataLength)
{
    int lrc = rPcb ^ rLen;
    int receivedCs = 0;
    int expectedCs = 0;
    NFC_DBG_MSG(KERN_INFO "receiveAndCheckChecksum - Enter\n");

    dataLength = dataLength - csSize;

    // compute the expected CS
    expectedCs = lrc ^ helperComputeLRC(data, 0, dataLength);

    // receive the T = 1 CS
    receive(filp,&checksum,csSize,C_TRANSMIT_NO_START_CONDITION);

    receivedCs = checksum[0];

    // compare the chechsums
    if (expectedCs ! = receivedCs)
    {
        NFC_DBG_MSG(KERN_INFO "Checksum error \n");
    }

    NFC_DBG_MSG(KERN_INFO "receiveAndCheckChecksum - Exit\n");
}
/**
```

* Basic send function which directly calls the spi bird wrapper function
 *
 * @param data
 *   the data to be send
 *
 */
```c
int send(struct file * filp,unsigned char * * data, unsigned char mode,int len)
{
    int count = 0;
    NFC_DBG_MSG(KERN_ALERT "send - Enter\n");
    NFC_DBG_MSG(KERN_ALERT "send - len = %d\n",len);

    // call to the spi bird wrapper
    count = p61_dev_write(filp, * data,len,0x00);

    if(count == 0) {
        NFC_ERR_MSG(KERN_ALERT "ERROR:Failed to send data to device\n");
        return -1;
    }
    return count;
}


int receive(struct file * filp,unsigned char * * data, int len, unsigned char mode)
{
    static int count_status;
    NFC_DBG_MSG(KERN_ALERT "receive - Enter\n");
    count_status = p61_dev_read(filp, * data,len,0x00);
    if(count_status == 0 && len != 0) {
        NFC_ERR_MSG(KERN_ALERT "ERROR:Failed to receive data from device\n");
        return -1;
    }

    NFC_DBG_MSG(KERN_ALERT "receive - Exit\n");

    return count_status;
}


static ssize_t p61_dev_write(struct file * filp, const char * buf,
```

```c
                    size_t count, loff_t *offset)
{
    int ret = -1;
    struct p61_dev  *p61_dev = NULL;
#if P61_DBG_LEVEL
    int i;
    for(i = 0;i <= count;i++){
        NFC_DBG_MSG("p61_dev_write : buf%d[0x%x]",i,buf[i]);
    }
#endif
    //char tmp[MAX_BUFFER_SIZE];
    NFC_ERR_MSG(KERN_ALERT "p61_dev_write - Enter\n");
    gpio_direction_output(P61_CS,0);
    p61_dev = filp->private_data;
    /* Write data */
    ret = spi_write(p61_dev->spi, buf, count);
    NFC_DBG_MSG ("spi_write status = %d\n",ret);
    if (ret < 0) {
        ret = -EIO;
    }

    NFC_DBG_MSG(KERN_ALERT "p61_dev_write - Exit\n");
    udelay(1000);
    gpio_direction_output(P61_CS,1);
    return count;
}


static ssize_t p61_dev_read(struct file *filp, char *buf,
                size_t count, loff_t *offset)
{
    int ret = -1;
    struct p61_dev *p61_dev = filp->private_data;
    NFC_DBG_MSG(KERN_ALERT "p61_dev_read - Enter \n");
    gpio_direction_output(P61_CS,0);
    mutex_lock(&p61_dev->read_mutex);
    NFC_ERR_MSG(KERN_ALERT "p61_dev_read - aquried mutex - calling spi_read \n");
    NFC_ERR_MSG(KERN_ALERT "p61_dev_read - test1 count [0x%x] \n",(int)count);
    /** Read data */
```

```c
#ifdef IRQ_ENABLE
    NFC_DBG_MSG(KERN_ALERT "*********** Test11 ***********\n");
    if (!gpio_get_value(p61_dev->irq_gpio)) {
        while (1) {
            NFC_DBG_MSG(KERN_ALERT "*********** Test1 ***********\n");
            NFC_DBG_MSG(" %s inside while(1) \n",__FUNCTION__);
            p61_dev->irq_enabled = true;
            enable_irq(p61_dev->spi->irq);
            ret = wait_event_interruptible(p61_dev->read_wq,!p61_dev->irq_enabled);
            p61_disable_irq(p61_dev);
            if (ret) {
                NFC_DBG_MSG("p61_disable_irq() : Failed\n");
                goto fails;
            }
            NFC_DBG_MSG(KERN_ALERT "*********** Test2 ***********\n");
            if (gpio_get_value(p61_dev->irq_gpio))
                break;

            NFC_DBG_MSG(" %s: spurious interrupt detected\n",__func__);
        }
    }

    NFC_DBG_MSG(KERN_ALERT "***********  gpio already high read data Test11 ***************\n");
#endif
    NFC_DBG_MSG(KERN_ALERT "*********** Test3 ***************\n");
    ret = spi_read(p61_dev->spi,(void *) buf, count);

#if P61_DBG_LEVEL
    int i;
    for(i = 0;i <= count;i++) {
        NFC_DBG_MSG(" == WAC == buf %d[0x%x]",i,buf[i]);
    }
#endif

    if(0 > ret) {
        NFC_ERR_MSG(KERN_ALERT "spi_read returns -1 \n");
        goto fails;
```

```
        }

        NFC_DBG_MSG(KERN_ALERT "Read ret % d \n",ret);
        mutex_unlock(&p61_dev - > read_mutex);
        gpio_direction_output(P61_CS,1);

        if (0 == ret) {
            ret = count;
        }

        return ret;

        fails:
        mutex_unlock(&p61_dev - > read_mutex);
        gpio_direction_output(P61_CS,1);
        return ret;
}


/**
 * This function is used to send a single T = 1 frame.
 *
 * @param data
 *              the data to be send
 * @param mode
 *              used to signal chaining
 *
 */
int sendFrame(struct file * filp, const char data[], char mode,int count)
{
    int count_status = 0;
    int len = count + headerSize + csSize;
    unsigned char * frame = NULL;
    NFC_ERR_MSG(KERN_INFO "sendFrame - Enter\n");
    frame = gSendframe;
    // update the send sequence counter of the terminal
    seqCounterTerm = (unsigned char)(seqCounterTerm ^ 1);

    // prepare the frame and send it
```

```c
        frame[0] = 0x00;
        frame[1] = (unsigned char)(mode |(unsigned char)(seqCounterTerm << 6));
        frame[2] = (unsigned char)(count);

        memcpy((frame + 3),data, count);

        frame[count + headerSize] = (unsigned char)helperComputeLRC(frame, 0, count + headerSize - 1);
        lastFrame = frame;
        lastFrameLen = len;
        count_status = send(filp,&frame, C_TRANSMIT_NORMAL_SPI_OPERATION,len);

        if(count_status == 0)
        {
            NFC_ERR_MSG(KERN_ALERT "ERROR:Failed to send device\n");
            return -1;
        }

        NFC_ERR_MSG(KERN_INFO   "sendFrame ret = %d - Exit\n", count_status);
        return count_status;
}
/**
    * Helper function to compute the LRC.
    *
    * @param data
    *           the data array
    * @param offset
    *           offset into the data array
    * @param length
    *           length value
    *
    */
unsigned char helperComputeLRC(unsigned char data[], int offset, int length)
{
    int LRC = 0;
    int i = 0;
    NFC_DBG_MSG(KERN_INFO "helperComputeLRC - Enter\n");
    for (i = offset; i < = length; i++)
```

```c
    {
        LRC = LRC ^ data[i];
    }
    NFC_DBG_MSG(KERN_INFO   "LRC Value is    %x \n",LRC);
    return (unsigned char)LRC;
}
/**
    * This function initializes the T = 1 module
    *
    */
void init()
{
    NFC_DBG_MSG(KERN_INFO   "init - Enter\n");
    apduBufferidx = 0;
    setAddress(0);
    setBitrate(100);

    NFC_DBG_MSG(KERN_INFO   "init - Exit\n");
}


void setAddress(short address)
{
    int stat = 0;
    NFC_DBG_MSG(KERN_INFO   "setAddress - Enter\n");
    stat = nativeSetAddress(address);

    if (stat != 0)
    {
        NFC_ERR_MSG(KERN_INFO   "set address failed.\n");
    }

    NFC_DBG_MSG(KERN_INFO   "setAddress - Exit\n");
}


void setBitrate(short bitrate)
{
    int stat = 0;
    NFC_DBG_MSG(KERN_INFO   "setBitrate - Enter\n");
```

```c
    stat = nativeSetBitrate(bitrate);

    if (stat != 0)
    {
        NFC_ERR_MSG(KERN_INFO  "set bitrate failed.\n");
    }

    NFC_DBG_MSG(KERN_INFO  "setBitrate - Exit\n");
}

int nativeSetAddress(short address)
{
    NFC_DBG_MSG(KERN_INFO  "nativeSetAddress - Enter\n");
    return 0;
}

int nativeSetBitrate(short bitrate)
{
    NFC_DBG_MSG(KERN_INFO  "nativeSetBitrate - Enter\n");

    return 0;
}

#ifdef IRQ_ENABLE
static void p61_disable_irq(struct p61_dev * p61_dev)
{
    unsigned long flags;

    NFC_DBG_MSG("Entry : %s\n", __FUNCTION__);
    spin_lock_irqsave(&p61_dev->irq_enabled_lock, flags);
    if (p61_dev->irq_enabled)
    {
        disable_irq_nosync(p61_dev->spi->irq);
        p61_dev->irq_enabled = false;
    }
    spin_unlock_irqrestore(&p61_dev->irq_enabled_lock, flags);
    NFC_DBG_MSG("Exit : %s\n", __FUNCTION__);
}
```

```c
static irqreturn_t p61_dev_irq_handler(int irq, void * dev_id)
{
    struct p61_dev * p61_dev = dev_id;

    NFC_DBG_MSG("Entry : % s\n", __FUNCTION__);
    p61_disable_irq(p61_dev);

    /* Wake up waiting readers */
    wake_up(&p61_dev -> read_wq);

    NFC_DBG_MSG("Exit : % s\n", __FUNCTION__);
    return IRQ_HANDLED;
}
#endif

#if 0
static int p61_enable_irq(struct p61_dev * p61_dev)
{
    int ret = -1;
    //unsigned int irq_flags;
    //NFC_DBG_MSG("p61_enable_irq Enter\n");
    NFC_DBG_MSG("Entry : % s\n", __FUNCTION__);
    ret = gpio_request( P61_IRQ, "p61 irq");
    if (ret < 0)
    {
        NFC_ERR_MSG("gpio request failed gpio = 0x % x\n", P61_IRQ);
        goto fail;
    }

    //add begin
    /* ret = request_irq(p61_dev -> spi -> irq, p61_dev_irq_handler, IRQF_TRIGGER_HIGH, p61_dev -> p61_device.name, p61_dev);
    NFC_DBG_MSG(" % s   name: % s   irq: % d\n", __FUNCTION__, p61_dev -> p61_device.name, p61_dev -> spi -> irq);
    if (ret)
    {
        NFC_ERR_MSG( "P61 request_irq failed\n");
        //goto err_request_irq_failed;
```

```
        }
     */
    //add end
    ret = gpio_direction_input(P61_IRQ);
    if (ret < 0)
    {
        NFC_ERR_MSG("gpio request failed gpio = 0x%x\n", P61_IRQ);
        goto fail_irq;
    }

    p61_dev->spi->irq = gpio_to_irq(P61_IRQ);

    if ( p61_dev->spi->irq < 0)
    {
        NFC_ERR_MSG("gpioi_to_irq request failed gpio = 0x%x\n", P61_IRQ);
        goto fail_irq;
    }

    NFC_DBG_MSG("Exit : %s\n", __FUNCTION__);

    return ret;
fail_gpio:
    gpio_free(P61_RST);
fail_irq:
    gpio_free(P61_IRQ);
fail:
    NFC_DBG_MSG("irq initialisation failed\n");
    return ret;
}
#endif
static inline void p61_set_data(struct spi_device *spi, void *data)
{
    dev_set_drvdata(&spi->dev, data);
}


static const struct file_operations p61_dev_fops = {
        .owner = THIS_MODULE,
        .llseek = no_llseek,
```

```
            .read = p61_dev_receiveData,
            .write = p61_dev_sendData,
            .open = p61_dev_open,
            .poll = p61_dev_poll,
            .unlocked_ioctl = p61_dev_ioctl,
            .compat_ioctl = p61_dev_ioctl,
    };

    static int p61_probe(struct spi_device * spi)
    {
        int ret = 0;
        struct p61_dev * p61_dev = NULL;
        unsigned int irq_flags;
        printk("P61 with irq without log Entry : %s\n", __FUNCTION__);

        NFC_DBG_MSG("chip select : %d, bus number = %d \n", spi -> chip_select, spi -> 
master -> bus_num);
        p61_dev = kzalloc(sizeof( * p61_dev), GFP_KERNEL);
        if (p61_dev == NULL) {
            NFC_ERR_MSG("failed to allocate memory for module data\n");
            ret = - ENOMEM;
            goto err_exit;
        }

    //      ret = gpio_request( P61_PWR, "p61 power");
    //      if (ret < 0) {
    //          NFC_ERR_MSG("p61 gpio reset request failed = 0x%x\n", P61_PWR);
    //          goto fail_gpio;
    //      }
    //      NFC_ERR_MSG("gpio_request returned = 0x%x\n", ret);
    //      ret = gpio_direction_output(P61_PWR,1);
    //      if (ret < 0) {
    //          NFC_ERR_MSG("p61 gpio rst request failed gpio = 0x%x\n", P61_PWR);
    //          goto fail_gpio;
    //      }
    //      NFC_ERR_MSG("gpio_direction_output returned = 0x%x\n", ret);

            ret = gpio_request( P61_CS, "p61 cs");
```

```c
    if (ret < 0) {
        NFC_ERR_MSG("p61 gpio cs request failed = 0x%x\n", P61_CS);
        goto fail_gpio;
    }
    NFC_ERR_MSG("gpio_request returned = 0x%x\n", ret);
    ret = gpio_direction_output(P61_CS,1);
    if (ret < 0) {
        NFC_ERR_MSG("p61 gpio cs request failed gpio = 0x%x\n", P61_CS);
        goto fail_gpio;
    }
    NFC_ERR_MSG("gpio_direction_output returned = 0x%x\n", ret);

    ret = gpio_request( P61_RST, "p61 reset");
    if (ret < 0) {
        NFC_ERR_MSG("p61 gpio reset request failed = 0x%x\n", P61_RST);
        goto fail_gpio;
    }

    NFC_ERR_MSG("gpio_request returned = 0x%x\n", ret);
    ret = gpio_direction_output(P61_RST,1);
    if (ret < 0) {
        NFC_ERR_MSG("p61 gpio rst request failed gpio = 0x%x\n", P61_RST);
        goto fail_gpio;
    }
    NFC_ERR_MSG("gpio_direction_output returned = 0x%x\n", ret);

#ifdef IRQ_ENABLE
    ret = gpio_request( P61_IRQ, "p61 irq");
    if (ret < 0) {
        NFC_ERR_MSG("p61 gpio request failed gpio = 0x%x\n", P61_IRQ);
        goto err_exit0;
    }
    ret = gpio_direction_input(P61_IRQ);
    if (ret < 0) {
        NFC_ERR_MSG("p61 gpio request failed gpio = 0x%x\n", P61_IRQ);
        goto err_exit0;
    }
#endif
```

```c
        spi -> bits_per_word = 8;
        spi -> mode = SPI_MODE_0;
        spi -> max_speed_hz = 7000000;//960000;//1000000;
        //spi -> chip_select = SPI_NO_CS;
        ret = spi_setup(spi);
        if'(ret < 0)
        {
            NFC_ERR_MSG("failed to do spi_setup()\n");
            goto err_exit0;
        }

        p61_dev -> spi = spi;
        p61_dev -> p61_device.minor = MISC_DYNAMIC_MINOR;
        p61_dev -> p61_device.name = "p61";
        p61_dev -> p61_device.fops = &p61_dev_fops;
        p61_dev -> p61_device.parent = &spi -> dev;

        p61_dev -> ven_gpio = P61_RST;
        gpio_set_value(P61_RST, 1);
        msleep(20);
        printk("p61_dev -> rst_gpio = %d\n ",P61_RST);
#ifdef IRQ_ENABLE
        p61_dev -> irq_gpio = P61_IRQ;
#endif

        p61_set_data(spi, p61_dev);
        /* init mutex and queues */
        init_waitqueue_head(&p61_dev -> read_wq);
        mutex_init(&p61_dev -> read_mutex);
        //spin_lock_init(&p61_dev -> irq_enabled_lock);
#ifdef IRQ_ENABLE
        spin_lock_init(&p61_dev -> irq_enabled_lock);
#endif

        ret = misc_register(&p61_dev -> p61_device);
        if (ret < 0) {
            NFC_ERR_MSG("misc_register failed! %d\n", ret);
            goto err_exit0;
```

```
    }
#ifdef IRQ_ENABLE
    p61_dev->spi->irq = gpio_to_irq(P61_IRQ);

    if (p61_dev->spi->irq < 0){
        NFC_ERR_MSG("gpio_to_irq request failed gpio = 0x%x\n", P61_IRQ);
        goto err_exit0;
    }
/* request irq.  the irq is set whenever the chip has data available
 * for reading.  it is cleared when all data has been read.
 */
    p61_dev->irq_enabled = true;
    irq_flags = IRQF_TRIGGER_RISING | IRQF_ONESHOT;

    ret = request_irq(p61_dev->spi->irq, p61_dev_irq_handler,
                    irq_flags, p61_dev->p61_device.name, p61_dev);
    if (ret) {
        NFC_ERR_MSG("request_irq failed\n");
        goto err_exit0;
    }
    p61_disable_irq(p61_dev);
#endif

    NFC_DBG_MSG("Exit : %s\n", __FUNCTION__);
    return ret;

//    err_exit1:
//    misc_deregister(&p61_dev->p61_device);
    err_exit0:
    mutex_destroy(&p61_dev->read_mutex);
    if(p61_dev != NULL)
    kfree(p61_dev);
    fail_gpio:
    gpio_free(P61_RST);
    err_exit:
    return ret;
}
static inline void * p61_get_data(const struct spi_device * spi)
```

```c
    {
        return dev_get_drvdata(&spi->dev);
    }

    static int p61_remove(struct spi_device * spi)
    {
        struct p61_dev * p61_dev = p61_get_data(spi);
        NFC_DBG_MSG("Entry : %s\n", __FUNCTION__);
        NFC_DBG_MSG(KERN_INFO " %s :: name : %s ", __FUNCTION__, p61_dev->p61_
device.name );

        free_irq(p61_dev->spi->irq, p61_dev);
        mutex_destroy(&p61_dev->read_mutex);
        misc_deregister(&p61_dev->p61_device);
        gpio_free(P61_IRQ);
        gpio_free(P61_RST);
        if(p61_dev != NULL)
            kfree(p61_dev);
        NFC_DBG_MSG("Exit : %s\n", __FUNCTION__);
        return 0;
    }
    static struct of_device_id p61_of_match_table[] = {
        { .compatible = "p61,nxp-nfc",},
        { },
    };
    static struct spi_driver p61_driver = {
            .driver = {
                    .name = "p61",
                    .bus = &spi_bus_type,
                    .owner = THIS_MODULE,
                    .of_match_table = p61_of_match_table,
            },
            .probe = p61_probe,
            .remove = p61_remove,
    //         .suspend = p61_suspend,
    //         .resume = p61_resume,
    };
    static int __init p61_dev_init(void)
```

```
{
    NFC_DBG_MSG("Entry : %s\n", __FUNCTION__);
    return spi_register_driver(&p61_driver);
    NFC_DBG_MSG("Exit : %s\n", __FUNCTION__);
}
module_init(p61_dev_init);

static void __exit p61_dev_exit(void)
{
    int val = -1;
    NFC_DBG_MSG("Entry : %s\n", __FUNCTION__);
    val = p61_dev_close();
    if(0 > val)
    {
        NFC_ERR_MSG("Falied free the memory : %s\n", __FUNCTION__);
    }
    spi_unregister_driver(&p61_driver);
    NFC_DBG_MSG("Exit : %s\n", __FUNCTION__);
}
module_exit(p61_dev_exit);

MODULE_AUTHOR("MANJUNATHA VENKATESH");
MODULE_DESCRIPTION("NFC P61 SPI driver");
MODULE_LICENSE("GPL");
```

4. NFC 硬件实体

NFC 硬件实体,这里一般就是指 NFC 射频前端控制器,主机端通过前面介绍的应用层、协议栈和驱动程序最终控制这个底层硬件。该硬件最主要的部分应该包含控制器和 RF 射频模块,前者负责处理和响应从主机端发下来 NCI 指令;后者负责与外部的 13.56 MHz 的 READR 模块或者卡片、标签进行数据通信和交换。当然,除了这两个最主要的部分外,还有例如数据和代码内存,与主机端的物理接口模块,支持 SWP SIM 接口的通道,另外还有一些如加解密处理的协处理单元等,如果需要支持嵌入式 eSE 的模块,那么还需另外增加硬件模块。如图 5-3 所示为一个普通的 NFC 硬件实体。

图 5-3 中的两个虚线区域,一个为嵌入式安全单元 eSE,另一个为天线。有些 NFC 硬件模块会在物理上把这两个部分封装在一起,这样就不需要额外地外挂 SE

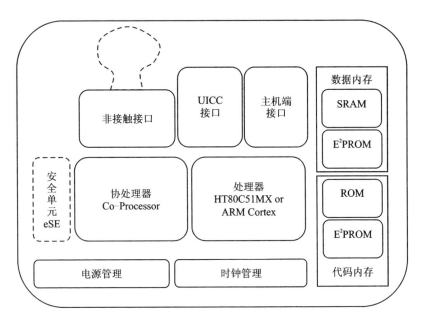

图 5-3 NFC 硬件实体

安全单元和天线电路，节约了 PCB 电路板的设计空间等；其缺点就是当有些设备终端的外部结构复杂时，内嵌天线就不方便进行适配调试了。

前面介绍了一个最基本的 NFC 近距离通信的 Android 最小系统，可以实现读/写器功能和点对点 P2P 功能，但是不包含卡模拟 CE 功能。其实对一个 NFC 系统来讲，主要应包括两大功能：第一，近距离通信功能，如读/写器功能和点对点 P2P 功能；第二，安全功能，如卡模拟 CE 功能，这个功能需要外加额外的安全单元 SE 载体才能实现，例如当下有一些场景有内嵌入式 eSE 载体、SWP SIM 方式的载体和 SWP SD 卡等。

5.1　基于 Android 系统的 NFC 应用支付框架

首先了解一下 Andorid 系统中 NFC 支付方式的参考硬件部分。如图 5-4 所示，以 Google 的 Nexus S 手机为例，手机包括两大部分：第一部分为 NFC 硬件模块，其中有 NFC 射频前端控制器和嵌入式 eSE 单元，还有支持 SWP SIM 卡的连接部分；第二部分为粘连在手机背壳上的 NFC 天线。Nexus S 手机的主机端通过 I^2C 总线以及其他的控制总线连接到 NFC 的硬件支付模块，NFC 的硬件支付模块再通过

5 基于 Android 系统的 NFC 实现框架

差分信号连接到背壳的天线上。

NFC 硬件支付模块的射频前端控制器部分使用的是恩智浦公司的 PN65 芯片，该芯片相当于把 NFC 射频前端控制器和嵌入式安全单元 eSE 封装在一起，并且支持 SWP 的接口，可以支持外接 SWP 的 SIM 卡或者 SD 卡。如图 5-4 所示，实际上该芯片只支持一路物理上的安全通道接口，嵌入式安全单元 eSE、SWP SIM 卡或者 SWP SD 卡，它们三组只能在同一时刻支持一路安全单元的工作，无法做到两路或者三路同时工作。如果需要物理硬件电路上支持 SWP SIM 和 SWP SD 卡，那么就需

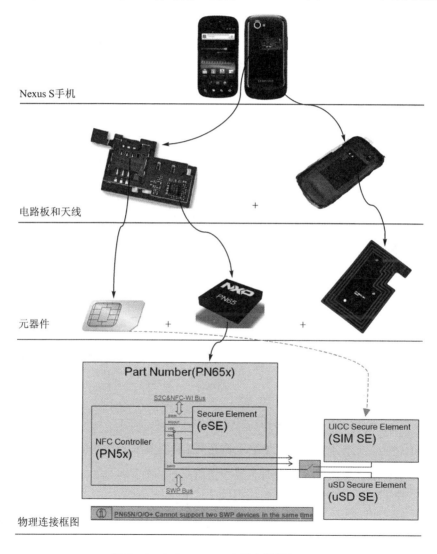

图 5-4 Android NFC 应用支付框架

要加入电子开关进行切换支持,需要同时支持多路安全单元接口的将在接下来的 PN80T 芯片中进行介绍。其中,NFC 射频前端控制器与嵌入式安全单元 eSE 之间使用的是恩智浦公司的私有连接协议 DWP 接口。

IEC/ISO 14443 规范定义频率范围为 13.56 MHz±7 kHz,理论上该规范的通信距离能达到十几厘米,但是由于手机的外形设计及系统的电气复杂性,以及天线尺寸和结构的限制,一般情况下手机或者穿戴设备采用的三种工作模式无法均衡达到这个通信的理论参考值。例如图 5-4 中的 Nexus S 手机,它的天线一边为塑料外壳,另一边为电池和电路板,因为金属原因会限制和衰减射频通信信号,所以一般的做法是在这种差分天线的后边贴上一层吸波材料(大部分为铁氧体材料),这样的设计虽然天线无法做到 360°的自由场通信空间,但至少对于塑料外壳那一侧的射频场辐射是没有问题的。对于当前在 NFC 的天线设计方面主要有两种方式:第一种为 Nexus S 天线的设计方式,NFC 射频前端控制器经过一个匹配网络,再通过差分方式接触到线圈缠绕的天线上;第二种为单端天线设计,顾名思义就是在之前的差分电路部分通过一个巴伦电路(Balun)的耦合设计,把之间的双端连接改成单端连接,另外一点接地即可。这种设计有几点好处:第一,可以复用设备外部的金属边框,降低物理天线和吸波材料的成本;第二,可以放置在手机的前额头处,让其在与外部 POS 机、卡片标签或 NFC P2P 设备通信时,有一个更大的通信射频空间场,这样用户的体验会比第一种差分方式更好。

下面将以恩智浦公司的 PN66T 和 PN80T 芯片为例,介绍其硬件方案在 Android 系统中的 NFC 应用支付框架。其中,PN66T 芯片实际上是由 PN548 和 P61 两个芯片封装在一起组成的,PN80T 则是由 PN553 和 P73 两个芯片封装在一起。可以这样理解,NFC 射频前端控制器和嵌入式安全单元 eSE 两个芯片与主机端的接口是可以独立进行工作的。

如图 5-5 所示,NFC 射频前端控制器与主机端的 REE 区域通过 I^2C 接口相连,外加硬件复位使能、固件升级、中断引脚连接;而嵌入式安全单元 eSE 通过 SPI 总线与主机端的 TEE 区域相连,市面上有些主机端的 TEE 部分只负责对数据总线的控制,GPIO 不支持在 TEE 中的控制,所以对于逻辑或者中断控制的部分还需要在 REE 区域进行控制,即在 REE 一侧做控制(即 Linux Driver)。例如中断控制引脚 IRQ 也不支持直接在 TEE 中响应,必须从 REE 中先做中断处理,然后再通知 TEE 中的 App Client 去做中断的相应处理程序,比如通过 SPI 总线发起对嵌入式安全单元 eSE 进行数据读/写操作等。如果支持营运商 UICC 支付方案，则需要将 SIM_VCC 和 SIM_SWIO 两个引脚连接到 SIM 卡槽上。

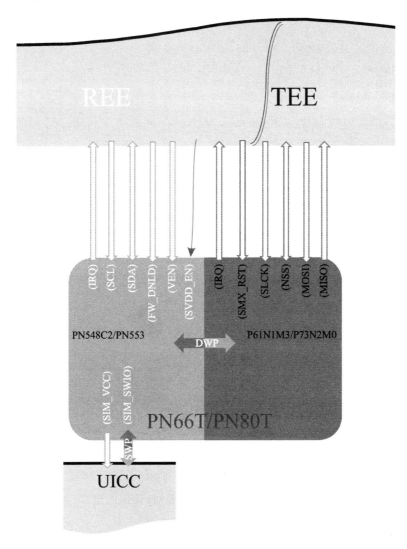

图 5-5　NFC 射频前端控制器与主机端连接示意图

图 5-5 所示为一个目前 NFC 支付模块的硬件连接设计，其中以 PN553 射频前端控制器为例，它是一颗基于 ARM Cortex-M0 的 CPU 内核，分别有 128 KB 的 ROM 和 28 KB 的 E^2PROM 的代码内存空间，8 KB 的 SRAM 和 4 KB 的 E^2PROM 的用户数据内存空间。与主机端接口可支持最高 3.4 Mbit/s 的 I^2C 被动模式和最高 7 Mbit/s 的 SPI 被动模式。

该 NFC 射频前端控制器支持的读/写器模式、卡模拟模式和 P2P 点对点模式的具体参数说明如下：

(1) 读/写器模式

- 支持 ISO/IEC 14443 A 标准的卡片；
- 支持 ISO/IEC 14443 B 标准的卡片；
- 支持 ISO/IEC 15693 标准的卡片；
- 恩智浦公司的 MIFARE 1k / 4k 卡片；
- 索尼公司的 FeliCa 卡片；
- NFC 论坛标准标签 1、2、3、4、5。

(2) 卡模拟模式

- ISO/IEC 14443 A 卡片；
- ISO/IEC 14443 B 卡片；
- ISO/IEC 14443 B'（仅支持通过 SWP 通道的 SE 安全单元模拟）；
- MIFARE 1k / 4k 卡片；
- MIFARE DESFire 卡片。

(3) 点对点模式

- 主动模式支持通信速率 106～424 kbit/s，支持发起端和接收端两种模式；
- 被动模式支持通信速率 106～424 kbit/s，支持发起端和接收端两种模式。

支持点对点数据共享的应用有：分享词条、网站链接和联系人名片；共享网页信息；音视频数据分享；应用程序分享。

恩智浦公司 BGA（Ball Grid Array）封装的 PN80T 芯片物理上由 64 个引脚构成，它与主机端接口的引脚信号说明如表 5-1 所列，其他的电源、地和射频网络等连接引脚需要查阅具体的数据手册，这里就不一一介绍了。

表 5-1 PN80T BGA 封装与主机信号接口说明

引 脚	引脚名称	引脚信号描述	信号说明
C3	HIF1	I2C_ADR0	配置 NFCC I^{2}C 器件从地址
D1	HIF2	I2C_ADR1	
E2	HIF4	NFC_I2C_SCL	NFCC 与主机端的 I^{2}C 总线接口
E1	HIF3	NFC_I2C_SDA	

续表 5-1

引脚	引脚名称	引脚信号描述	信号说明
E3	IRQ	NFC_IRQ	NFCC 输出给主机端的中断信号
H1	VEN	NFC_ENABLE	NFCC 的复位控制引脚
D3	DWL_REQ	NFC_DWL_REQ	NFCC 的固件升级控制引脚
B2	PWR_REQ	SVDD_PWR_REQ	NFCC 给 eSE 供电的控制信号
G4	ESE_IRQ	ESE_IRQ	eSE 输出给主机端的中断信号
E5	ESE_SPI_CS_N	ESE_SPI_CS_N	eSE 与主机端的 SPI 总线接口
F5	ESE_SPI_MOSI	ESE_SPI_MOSI	
C4	ESE_SPI_MISO	ESE_SPI_MISO	
D5	SMX_CLK	ESE_SPI_CLK	
F6	SMX_RST_N	ESE_RSTN	eSE 的复位控制引脚

其中 PN80T BGA 封装与主机信号接口中关于 I^2C 通信速率的设置,最高可设置为 3.4 MHz,最低设置推荐不要低于 100 kHz(因为在某些应用场景,由于有交易时间等问题,例如在主机端处于休眠状况下,与外部 POS 机进行交易,主机端从唤醒再到把 NFC 射频前端控制器 I^2C 总线上的数据读走,如果通信速率过低,那么就会造成交易超时等问题)。在最高推荐速率与最低推荐速率之间,可以设置其他的通信速率,具体设置与主机端和具体电路板的信号设计等有关,如表 5-2 所列为 I^2C 推荐的速率设置表。

PN80T 的 I^2C 地址设定规则为,在 E^2PROM 的某一个内存中会有两个参数 0x9E 和 0x9C 来定义 I^2C 器件地址的组成规则。数据示例如下:

27 36 9E/9C B4 9A 01 04 00 00 00 00 00 00 00 00 00

0x9C:the I^2C address is defined by the electrical logic levels on ADR0 and ADR1 pins。

地址由 I2C_ADR0 和 I2C_ADR1 引脚的逻辑电平决定。

0x9E:the ADR0=0 whatever the electrical logic level on the ADR1 pin。

I2C_ADR0 引脚的逻辑电平不做参考,一直为 0,只需要参考 I2C_ADR1 的逻辑电平状态。

一般出厂的芯片都会默认设置 0x9c,也就是说 I^2C 的器件地址由"0 1 0 1 0 I2C_ADR0 I2C_ADR1"组成,由这两个引脚决定。一般硬件设计同一个 I^2C 总线时,如

果 0x28/0x50 没有 I²C 物理设备冲突问题，则推荐这两个引脚(C3,D1)直接接地，即实际的物理地址为 0x50，但是在驱动代码中会经过算法"outb_p(((addr & 0x7f) << 1) | (read_write & 0x01)"转换，所以这个地址值在驱动中就应该设置为 0x28。如表 5-3 所列为 PN80T 的 I²C 地址编码规则。

表 5-2 I²C 的推荐速率设置表

SCL 速率档位	DTS 代码示例	说 明

Standard mode (100 kHz)	i2c1：i2c@44e0b000 {
	clock-frequency=<100000>;
	tps：tps@2d {
	reg=<0x2d>;
	};
		不要低于 100 kHz

Fast mode (400 kHz)	i2c1：i2c@44e0b000 {
	clock-frequency=<400000>;
	tps：tps@2d {
	reg=<0x2d>;
	};
		推荐

High-speed mode (3.4 MHz)	i2c1：i2c@44e0b000 {
	clock-frequency=<3400000>;
	tps：tps@2d {
	reg=<0x2d>;
	};
		最高速率设置

表 5-3 PN80T 的 I²C 地址编码

0 1 0 1 0 I2C_ADR0 I2C_ADR1 R/W	驱动代码设定的地址
0 1 0 1 0 0 0 x(0x50/0x51)	0x28
0 1 0 1 0 0 1 x(0x52/0x53)	0x29
0 1 0 1 0 1 0 x(0x54/0x55)	0x2a
0 1 0 1 0 1 1 x(0x56/0x57)	0x2b

PN80T 芯片中的 eSE 芯片 P73 为一颗基于 ARM SC300 的 CPU 内核，与主机端连接支持最高 10 MHz 的 ISO/IEC 7816 通信接口和最高 10 Mbit/s 波特率的 SPI 被动模式，支持 PKC、RSA、DES、3DES、ECC、AES 等硬件加密算法，支持高速 8 位、

5 基于 Android 系统的 NFC 实现框架

16 位和 32 位的 CRC 硬件引擎,支持产生真随机数。该芯片一旦正常工作,与其通信的数据格式为标准的 ISO/IEC 7816 APDU 数据包,如图 5-5 中主机端的 TEE 直接通过 SPI 总线发送 APDU 数据即可。具体关于安全域的生命周期管理,需要查看 GP（Gloab Platform）规范,APDU 命令格式说明需要查看对应的 Applet 数据接口或规范。

下面简单介绍一下通用的 APDU 数据格式,APDU 分为两种类型:命令型的 APDU 和响应型的 APDU,即命令和返回值有不同的格式。

命令型的 APDU 主要由指令类别、指令、参数、长度及数据体构成,指令类别字段（CLA,Class）用来标识命令的各种分类,比如属于哪个规范内的定义,包含消息是否加密和发给哪个逻辑通道等;指令代码字段（INS,Instruction）是具体的操作指令,比如 SELECT AID、INSTALL APPLET、UPDATE RECORD 等;指令参数字段（P1,P2,Parameter 1,2）是指令的参数,进一步限定了指令的具体行为;指令长度字段（Lc,Length）为后面命令数据体的长度,比如 SELECT AID 指令,指令长度就包含了后面的数据体中 AID 的一系列字节,从而限定了这次发送的数据体的最大长度;命令数据体字段（Data Field）为命令所要传递过去的实际数据值,再以 SELECT AID 指令为例,那么这里就要选择真实的 AID 数值;期待返回数据的长度字段（Le,Length）是指发送给命令接收者,期待返回数据的最大长度。如表 5-4 所列为 IEC/ISO 7816-4 命令型 APDU 数据组成格式。

表 5-4　IEC/ISO 7816-4 命令型 APDU 数据组成格式

字 段	长度参考/字节	说 明	强制/可选项
CLA	1	指令的类别	强制项
INS	1	指令代码	强制项
P1	1	参数 1	强制项
P2	1	参数 2	强制项
Lc	0,1 or 3	命令数据字节数	可选项
Data	等于 Lc 的长度	命令数据	可选项
Le	0~3	期待返回数据的最大字节数	可选项

响应型的 APDU 由数据体字段（Data Field）和状态字段（SW1,2,Status Word 1,2）组成,前者为命令 APDU 的响应返回数据,后者为响应和处理命令后自身的状态报告。如表 5-5 所列为 IEC/ISO 7816-4 响应型 APDU 数据组成格式。

表 5 – 5　IEC/ISO 7816 – 4 响应型 APDU 数据组成格式

字　段	长度参考/字节	说　明	强制/可选项
Data	Lr	响应数据	可选项
SW1,2	2	响应命令后的状态	强制项

表中的状态字的参考代码数组如下：

```
typedef struct
{
    uint8_t status_word1;
    uint8_t status_word2;
    char p_string[0xff];//255 characters
}p_string_table;

p_string_table p_table[] =
{
    /* Normal processing */
    {0x90,0x00,"No error"},
    {0x61,0x00/* 0x00～0xff */,"Normal processing,< SW2 > number of data bytes still avaliable"},

    /* Warning processing */
    {0x62,0x00,"No information given"},
    {0x62,0x02/* 0x02～0x80 */,"should retrieve and for which the < SW2 > possibly expects a response"},
    {0x62,0x81,"Part of returned data maybe corrupted"},
    {0x62,0x82,"End of file or record reached before reading Ne bytes"},
    {0x62,0x83,"Selected file deactivated"},
    {0x62,0x84,"File control information not formatted"},
    {0x62,0x85,"Selected file in termination state"},
    {0x62,0x86,"No input data available from a sensor on the card"},

    {0x63,0x00,"No information given"},
    {0x63,0x81,"File filled up by the last write"},
    {0x63,0xc0/* 0xc0～0xcf */,"SW2|0x0f Counter Warning"},

    /* Execution error */
    {0x64,0x00,"Execution error"},
```

{0x64,0x01,"Immediate response required by the card"},

{0x64,0x02/ * 0x02~0x80 * /,"should retrieve and for which the < SW2 > possibly expects a response"},

{0x65,0x00,"No information given"},
{0x65,0x81,"Memory failure"},

{0x66,0x00/ * 0x00~0xff * /,"Security related issues"},

/ * Checking error * /
{0x67,0x00,"Wrong length"},

{0x68,0x00,"No information given"},
{0x68,0x81,"Logical channel not supported"},
{0x68,0x82,"Secure messaging not supported"},
{0x68,0x83,"Last command of the chain expected"},
{0x68,0x84,"Command chaining nor supported"},

{0x69,0x00,"No information given"},
{0x69,0x81,"Command incompatible with file structure"},
{0x69,0x82,"Security status not satisfied"},
{0x69,0x83,"Authentication method blocked"},
{0x69,0x84,"Referce data not usable"},
{0x69,0x85,"Conditions of use not satisfied"},
{0x69,0x86,"Command not allowed(no current EF)"},
{0x69,0x87,"Expected secure messaging data objects missing"},
{0x69,0x88,"Incorrect secure messaging data objects"},

{0x6a,0x00,"No information given"},
{0x6a,0x80,"Incorrect parameters in the command data field"},
{0x6a,0x81,"Function not supported"},
{0x6a,0x82,"File or application not found"},
{0x6a,0x83,"Record not found"},
{0x6a,0x84,"Not enough memory space in the file"},
{0x6a,0x85,"Nc inconsistent with TLV structure"},
{0x6a,0x86,"Incorrect parameters P1 - P2"},
{0x6a,0x87,"Nc inconsistent with parameters P1 - P2"},
{0x6a,0x88,"Reference data or reference data not found"},

            {0x6a,0x89,"File already exists"},
            {0x6a,0x8a,"DF name already exists"},

            {0x6b,0x00,"Wrong parameters P1 – P2"},

            {0x6c,0x00/ * 0x00 ~ 0xff * /,"Process aborted,'SW2' number of data bytes still avaliable"},

            {0x6d,0x00,"Instruction code not supportd or invaild"},
            {0x6e,0x00,"Class not supported"},

            {0x6f,0x00,"No precise diagnosis"},
        };

APDU 的命令集的可扩展性很强,也就是说最准确的定义应该由其具体的应用程序进行确定。因为各个行业,比如交通系统、金融系统等都会对自己行业相关的卡片应用定义相应的规范,其中涉及到很多规范。默认情况下,大家都参考 IEC/ISO 7816 – 4 和 Global Platform 两个基础规范,少数情况下有些行业应用会有自己的修改或者增加私有命令格式,这时就需要考虑对应的该程序曾具体参考过哪些标准和规范等,本书对此部分不做过多介绍。

上面重点介绍了 Android 系统中 NFC 支付系统的硬件框架,如图 5 – 6 所示为包含嵌入式 eSE 安全单元、NFC 射频前端控制器、UICC、基带芯片和 SWP SD 卡的一个支付软件系统的框架说明。

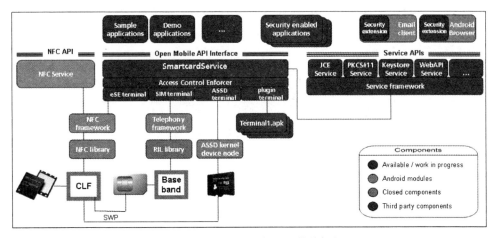

图 5 – 6 Android 系统 NFC 软件框架

5 基于 Android 系统的 NFC 实现框架

Android 系统默认 NFC 软件是跑在主机端上的,最下方为 5 个硬件实体,从左到右分别为嵌入式 eSE 安全单元、NFC 射频前端控制器(CLF,Contactless Front-end)、UICC、基带芯片和 SWP SD 卡。这里我们分成 5 种应用场景进行解释,分别为嵌入式安全单元 eSE 方式、SWP SIM 方式、SWP SD 方式、HCE 主机端模拟方式和程序端嵌入式方式。

(1) 嵌入式安全单元 eSE 方式

前面专门介绍过 NFC API 接口主要是给上层第三方提供应用程序,例如 NFC 的读卡器模式和 P2P 模拟的应用调用接口等。它们通过 NFC Framework 和 NFC 协议栈对 NFC 射频前端控制器进行控制和通信。目前,嵌入式安全单元 eSE 方式的卡模拟方式主要有如下几种工作通路:

> 对于有系统权限的应用,例如 Huawei Pay、Mi Pay 等应用,可以通过 NFC Service 的模式直接对 eSE 发起访问,只是需要把该系统应用 APK 的 android: signature 签名加入到 nfcse_access.xml 和 nfcee_access.xml 文件中。

> 通过 Open Mobile API 的这层方式,再调用 NFC Framework 和 NFC 协议栈,对 NFC 射频前端控制器以及嵌入式 eSE 安全单元进行访问和通信,这种方式对于第三方应用程序希望去访问嵌入式 eSE 安全单元是比较好的方式,也是目前唯一的一个通路。第三方开发者在开发使用 SE 安全单元时,需要在 IDE 环境下集成和调用 SmartcardService 的 API 接口,并且当应用程序开发完成后需要对应用 APK 算一次 SHA1 签名,这个 APK 的 SHA1 值需要在应用部署前把该值写入到 SE 安全单元与 Access Control Enforcer 相对应的 ARAC applet 中,这样这个第三方的应用程序才有可能访问到 SE 安全单元。

> 上面两种方式的物理通道其实走的都是 I^2C 的物理通道,且主机端通过 REE 的权限进行访问 NFC 射频前端控制器,控制器再通过内部连接通道发起与嵌入式安全单元 eSE 之间的通信。现在主流的基于嵌入式安全单元 eSE 的 NFC 支付系统,一般主机端与嵌入式安全单元 eSE 之间,会通过 TEE 权限走 UART 或者 SPI 通道,这样通过 TEE 端有一个专门管理的可信赖的应用程序与嵌入式安全单元 eSE 进行通信。这种模式目前只有拥有系统级权限的支付程序才能做到,第三方程序如果想通过 TEE 的可信赖的应用程序去调用安全单元,需要做特别的证书提前预置和进行系统配置。

> 前面的三种方式主要应用于开卡、充值和销卡等，对于与外部 POS 机交易的接口首先是通过 NFC 射频前端控制器与外部天线进行数据交换，然后通过 NFC 射频前端控制器与嵌入式安全单元 eSE 的内部接口进行数据单元交易。

上面介绍的 Android 应用程序 APK 的 SHA1 和 android:signature 签名，可以通过如下参考代码 SignPicker.java 进行 APK 的数据提取。

```java
package xiaohua;

import java.io.FileWriter;
import java.io.IOException;
import java.io.InputStream;
import java.lang.ref.WeakReference;
import java.security.MessageDigest;
import java.security.NoSuchAlgorithmException;
import java.security.Signature;
import java.security.cert.*;
import java.util.Enumeration;
import java.util.jar.JarEntry;
import java.util.jar.JarFile;
import java.util.logging.Level;
import java.util.logging.Logger;

public class SignPicker {

    private static final Object mSync = new Object();
    private static WeakReference < byte[] > mReadBuffer;
    private static String apk_filePath;
    private static String apk_fileName;
    private static String sha1 = null;

    public static void main(String[] args) {
        if (args.length < 1) {
            System.out.println("Version:1.0");
            System.out.println("OS:" + System.getProperties().getProperty("os.name"));
            System.out.println("Usage:\tjava - jar signpicker.jar < apk/jar >");
```

```
            System.out.println("Scope:\tA utility to dig APK signature");
            System.out
                    .println("\tIf you meet some problems, please contact xiaohua.
wang@nxp.com");
            System.out
                    .println("\tCopyright (c) 2011 - 2015, NXP Semiconductors
N.V.");
            System.out
                    .println("Steps:\n\t(1) Put the tool < signpicker.jar > and
file < xxx.apk > under a path\n\t(2) Find your computer 'JAVA_HOME' environment\n\t(3) Open
'cmd' and input 'java - jar signpicker.jar < apk/jar > '");
            System.out
                    .println("Example:\n\tC:\\ > java - jar signpicker.jar JRCPTest_
Spi.apk\n\tC:\\Program Files\\Java\\jre7\\bin > java - jar signpicker.jar LTSM.apk");
            System.out
                    .println("\tC:\\Program Files (x86)\\Java\\jre6\\bin > java -
jar signpicker.jar NfcJRCP.apk");
            System.exit(-1);
        }

        // System.out.println(args[0]);

        String mArchiveSourcePath = args[0];
        int lastIndex = mArchiveSourcePath.lastIndexOf("\\");
        if (lastIndex == -1) {
            apk_filePath = System.getProperty("user.dir");
            apk_fileName = args[0];
        } else {
            apk_filePath = mArchiveSourcePath.substring(0, lastIndex);
            apk_fileName = mArchiveSourcePath.substring(++lastIndex);
        }

        WeakReference < byte[] > readBufferRef;
        byte[] readBuffer = null;
        synchronized (mSync) {
            readBufferRef = mReadBuffer;
            if (readBufferRef != null) {
                mReadBuffer = null;
```

```java
            readBuffer = readBufferRef.get();
        }
        if (readBuffer == null) {
            readBuffer = new byte[8192];
            readBufferRef = new WeakReference < byte[] > (readBuffer);
        }
    }

    try {
        JarFile jarFile = new JarFile(mArchiveSourcePath);
        java.security.cert.Certificate[] certs = null;

        Enumeration entries = jarFile.entries();
        while (entries.hasMoreElements()) {
            JarEntry je = (JarEntry) entries.nextElement();
            if (je.isDirectory()) {
                continue;
            }
            if (je.getName().startsWith("META - INF/")) {
                continue;
            }
            java.security.cert.Certificate[] localCerts = loadCertificates(
                    jarFile, je, readBuffer);
            if (false) {
                System.out.println("File " + mArchiveSourcePath + " entry " +
                        je.getName() + ": certs = " + certs + " (" +
                        (certs != null ? certs.length : 0) + ")");
            }
            if (localCerts == null) {
                System.err.println("Package has no certificates at entry " +
                        je.getName() + "; ignoring!");
                jarFile.close();
                return;
            } else if (certs == null) {
                certs = localCerts;
            } else {
                // Ensure all certificates match.
                for (int i = 0; i < certs.length; i ++) {
```

```
                    boolean found = false;
                    for (int j = 0; j < localCerts.length; j++) {
                        if (certs[i] != null
                                && certs[i].equals(localCerts[j])) {
                            found = true;
                            break;
                        }
                    }
                    if (!found || certs.length != localCerts.length) {
                        System.err.println("Package has mismatched
certificates at entry " + je.getName() + "; ignoring!");
                        jarFile.close();
                        return; // false
                    }
                }
            }
        }

        jarFile.close();

        synchronized (mSync) {
            mReadBuffer = readBufferRef;
        }

//              System.out.println("\n     <!-- " + fileName + " Cert#:" + i
//                       + "  Type:" + certs[i].getType()
//                       /* + "\tPublic key: " + certs[i].getPublicKey() */
//                       + " Hash code:" + certs[i].hashCode() + "/"
//                       + Integer.toHexString(certs[i].hashCode())
//                       + " SHA1:" + sha1 + " -->"
//                       + "\n    <signer android:signature = \"" + charSig
//                       + "\" />\n");

            if (certs != null && certs.length > 0) {
                final int N = certs.length;
                StringBuilder builder = new StringBuilder();
                builder.append("<?xml version = \"1.0\" encoding = \"utf-8\"? >
\n");
```

```
                builder.append("  < resources xmlns:xliff = \"urn:oasis:names:tc:
xliff:document:1.2\" > \n");

                System.out.println("\n\r");
                System.out.println(" < ? xml version = \"1.0\" encoding = \"utf - 8
\"? >");
                System.out.println(" < resources xmlns:xliff = \"urn:oasis:names:
tc:xliff:document:1.2\" > ");
                for (int i = 0; i < N; i++) {
                    String charSig = new String(toChars(certs[i].getEncoded()));

                    try {
                        sha1 = getThumbPrint(certs[i]);
                    } catch (NoSuchAlgorithmException e) {
                        // TODO Auto - generated catch block
                        e.printStackTrace();
                    }

                    System.out.println("\n    < ! -- " + apk_fileName + " Cert #:" + i
                            + "  Type:" + certs[i].getType()
                            /* + "\tPublic key: " + certs[i].getPublicKey()
                            + " Hash code:" + certs[i].hashCode() + "/"
                            + Integer.toHexString(certs[i].hashCode()) */
                            + " SHA1:" + sha1 + " -- >"
                            + "\n    < signer android:signature = \"" + charSig
                            + "\" / > \n");

                    builder.append("\n    < ! -- " + apk_fileName + " Cert #:" + i
                            + "  Type:" + certs[i].getType()
                            + " SHA1:" + sha1 + " -- >"
                            + "\n    < signer android:signature = \"" + charSig
                            + "\" / > \n\n");
                }
                System.out.println(" < /resources > ");
                System.out.println("\n\r");

                builder.append(" < /resources > \n");
                builder.append("\n\r");
```

```
                writeMethod(builder.toString());

            } else {
                System.err.println("Package has no certificates; ignoring!");
                return;
            }
        } catch (CertificateEncodingException ex) {
            Logger.getLogger(SignPicker.class.getName()).log(Level.SEVERE,
                    null, ex);
        } catch (IOException e) {
            System.err.println("Exception reading " + mArchiveSourcePath + "\n"
                    + c);
            return;
        } catch (RuntimeException e) {
            System.err.println("Exception reading " + mArchiveSourcePath + "\n"
                    + e);
            return;
        }
    }

    private static char[] toChars(byte[] mSignature) {
        byte[] sig = mSignature;
        final int N = sig.length;
        final int N2 = N * 2;
        char[] text = new char[N2];

        for (int j = 0; j < N; j++) {
            byte v = sig[j];
            int d = (v >> 4) & 0xf;
            text[j * 2] = (char) (d >= 10 ? ('A' + d - 10) : ('0' + d));
            d = v & 0xf;
            text[j * 2 + 1] = (char) (d >= 10 ? ('A' + d - 10) : ('0' + d));
        }

        return text;
    }

    private static java.security.cert.Certificate[] loadCertificates(
```

```java
                    JarFile jarFile, JarEntry je, byte[] readBuffer) {
            try {
                // We must read the stream for the JarEntry to retrieve
                // its certificates.
                InputStream is = jarFile.getInputStream(je);
                while (is.read(readBuffer, 0, readBuffer.length) != -1) {
                    // not using
                }
                is.close();

                return (java.security.cert.Certificate[]) (je != null ? je
                        .getCertificates() : null);
            } catch (IOException e) {
                System.err.println("Exception reading " + je.getName() + " in "
                        + jarFile.getName() + ": " + e);
            }
            return null;
        }


        public static void writeMethod(String s) {
            String fileName = apk_filePath + System.getProperties().getProperty("file.separator") + "nfcse_access.xml";
            String fileName_2 = apk_filePath + System.getProperties().getProperty("file.separator") + "nfcee_access.xml";
            String fileName_3 = apk_filePath + System.getProperties().getProperty("file.separator") + "ara_c.jcsh";
            try {
                FileWriter writer = new FileWriter(fileName);
                writer.write(s);
                writer.close();
                FileWriter writer_2 = new FileWriter(fileName_2);
                writer_2.write(s);
                writer_2.close();
                FileWriter writer_3 = new FileWriter(fileName_3);
                writer_3.write("# version 1.0, Write the Applet AID bind APK SHA1 into ARAC." +
                        "\n\n/echo \"script start! \"" +
                        "\n/atr" +
```

```
                "\n/card" +
                "\n/select A00000015141434C00" +
                "\n\n#TODO HERE!!! INPUT KEYS FOR ARA-C SCP" +
                "\n#/set-var KVN 34" +
                "\n#set-key ${KVN}/1/DES-ECB/404142434445464748494a4b4c4d4e4f
${KVN}/2/DES-ECB/404142434445464748494a4b4c4d4e4f  ${KVN}/3/DES-ECB/
404142434445464748494a4b4c4d4e4f" +
                "\n\ni-u ${KVN}" +
                "\ne-a mac" +
                "\n\n#TODO HERE!!! INPUT APPLET AID" +
                "\n#/set-var APLT 01020304050607" +
                "\nsend 80E29000 # (F02BE229E11F4F # ($ {APLT}) C114" + sha1 + "
E306D00101D10101)\n\n" +
                "\n/echo \"Things done! \"\n"
                );
                writer_3.close();
            } catch (IOException e) {
                e.printStackTrace();
            }
            System.out
                    .println("\n\n【1】APK ACCESS SE VIA NFC-EE\n================
=====================\n(a)Patch above the signature into phone \n    /system/
etc/nfcse_access & nfcee_access.xml\n(b)Or replace the file with below XML:"
                            + "\n    "
                            + fileName
                            + "\n    "
                            + fileName_2
                            + "\n(c)Ensure " + apk_fileName + " AndroidManifest.xml
have:" +
                            "\n    <uses-permission android:name=\"com.android.nfc.
permission.NFCEE_ADMIN\"></uses-permission>" +
                            "\n    <uses-permission android:name=\"android.permis-
sion.NFC\"></uses-permission>" +
                            "\n    <uses-permission android:name=\"android.permis-
sion.WRITE_SECURE_SETTINGS\"></uses-permission>" +
                            "\n    <uses-library" +
                            "\n    android:name=\"com.android.nfc_extras\"" +
                            "\n    android:required=\"true\"/>" +
```

```
                              "\n\n【2】APK ACCESS SE VIA OMA with ARA\n = = = = = = = = = = = = =
 = = = = = = = = = = = = = = = = = = = = = = = =" +
                              "\nModify the \"set - key\" and \"applet aid\"item in file " +
fileName_3
                              + "\n\nIf you meet some problems, please contact xiaohua.
wang@nxp.com");
                }

            public static String getThumbPrint(Certificate certs)
                    throws NoSuchAlgorithmException, CertificateEncodingException {
                MessageDigest md = MessageDigest.getInstance("SHA - 1");
                byte der[] = certs.getEncoded();
                md.update(der);
                byte digest[] = md.digest();
                return hexify(digest);

            }

            public static String hexify(byte bytes[]) {
//            char[] hexDigits = { '0', '1', '2', '3', '4', '5', '6', '7', '8', '9',
//                                 'A', 'B', 'C', 'D', 'D', 'F' };
                String str = null;

                StringBuffer buf = new StringBuffer(bytes.length * 2);

                for (int i = 0; i < bytes.length; ++i) {
//                buf.append(hexDigits[(bytes[i] & 0xf0) >> 4]);
//                buf.append(hexDigits[bytes[i] & 0x0f]);
                    buf.append(Integer.toString((bytes[i] & 0xff) + 0x100, 16).substring
(1));
                }
                str = buf.toString();
                return str.toUpperCase();
            }
        }
```

（2）SWP SIM 方式

SWP SIM 其实有两种工作模式：第一种为在非接触交易时，通过 NFC 射频前端

5 基于 Android 系统的 NFC 实现框架

芯片的天线与外界 POS 机进行通信,然后再通过 SWP 的物理通道与外部的 SIM 卡进行交易指令的交互;第二种为当要对 SWP SIM 卡进行开卡、充值或者销卡时,需要通过基带芯片的 IEC/ISO 7816 的接触通道进行工作。

- 在 SWP SIM 初始化完成后,NFC 射频前端控制器会把相应的 SWP SIM 的非接触参数记录到控制器的内存中,当有外部 POS 机靠近时,天线接收到场强信号,NFC 射频前端控制器与外部 POS 机建立握手通信,并且把一些内存中已经记录的 SWP SIM 的非接触参数,例如 UID、ATQA、B、SAK 等参数告知外部 POS 机,整个过程为全部的 IEC/ISO 14443 - 1,2,3 的协议握手,在进行到 APDU 数据交换层时,NFC 射频前端控制器会通过物理 SWP 通道把交易数据转发到 SWP SIM,从而完成整个非接触交易。
- 还有一种应用场景为开卡、充值或者销卡,这时上层支付应用程序需要通过 Open Mobile API 接口,调用 SmartcardServer 的 SDK,再通过 RIL 接口或者库文件调用基带芯片,然后通过 IEC/ISO 7816 的接触通道实现对 SIM 的业务层面的访问。

(3) SWP SD 方式

SWP SD 方式其实和 SWP SIM 方式非常相似,唯一的差别就是这种 SD 算是一非主流标准的 SD 卡,它需要额外设计增加电源和 SWP 的物理接口。SWP SD 也有两种工作模式:第一种为在非接触交易时,通过 NFC 射频前端芯片的天线与外界 POS 机进行通信,然后再通过 SWP 的物理通道与外部的 SD 卡进行交易指令的交互;第二种为当要对 SWP SD 卡进行开卡、充值或者销卡时,需要通过基带芯片的 IEC/ISO 7816 的接触通道进行工作。

- 在 SWP SD 初始化完成后,NFC 射频前端控制器会把相应的 SWP SD 卡的非接触参数记录到控制器的内存中,当有外部 POS 机靠近时,天线接收到场强信号,NFC 射频前端控制器与外部 POS 机建立握手通信,并且把一些内存中已经记录的 SWP SD 的非接触参数,例如 UID、ATQA、B、SAK 等参数告知外部 POS 机,整个过程为全部的 IEC/ISO 14443 - 1,2,3 的协议握手,在进行到 APDU 数据交换层时,NFC 射频前端控制器会通过物理 SWP 通道把交易数据转发到 SWP SD,从而完成整个非接触交易。
- 还有一种应用场景为开卡、充值或者销卡,这时上层支付应用程序需要通过 Open Mobile API 接口,调用 SmartcardServer 的 SDK,再通过 RIL 接口或者库文件调用基带芯片,然后通过 IEC/ISO 7816 的接触通道实现对 SD 卡的业务层面的访问。

(4) HCE 主机端模拟方式

这种模式主要的实现原理就是使用软件的方式把 SE 安全单元交易的 APDU 数据路由到 HCE 的主机端软件上,由 HCE 软件完成对外部 POS 机发起的 APDU 指令交换。

Android 提供了一个基类 HostApduService,首先在开发 HCE 应用时需要继承这个基类,另外还有一些资源文件需要进行相关配置。下面为一个 HCE 最基本的源代码示例程序。

源文件 MainActivity.java 如下:

```
package com.example.hce;

import android.nfc.cardemulation.HostApduService;
import android.os.Bundle;
import android.app.Activity;
import android.util.Log;
import android.view.Menu;
import android.widget.Toast;

//public class MainActivity extends Activity {
//
//      @Override
//      protected void onCreate(Bundle savedInstanceState) {
//          super.onCreate(savedInstanceState);
//          setContentView(R.layout.activity_main);
//      }
//
//      @Override
//      public boolean onCreateOptionsMenu(Menu menu) {
//          // Inflate the menu; this adds items to the action bar if it is present.
//          getMenuInflater().inflate(R.menu.main, menu);
//          return true;
//      }
//
//}
```

5 基于 Android 系统的 NFC 实现框架

```java
public class MainActivity extends HostApduService {

    @Override
    public byte[] processCommandApdu(byte[] apdu, Bundle extras) {
        byte[] responseApdu = new byte[] { (byte)0x6F, (byte)0x10, (byte)0x84, (byte)0x08, (byte)0xA0, (byte)0x020,(byte) 0x20, (byte)0x20, (byte)0x03, (byte)0x20, (byte)0x10,(byte) 0x20, (byte)0xA5, (byte)0x04,(byte) 0x9F, (byte)0x65, (byte)0x01, (byte)0XFF, (byte)0x90, (byte)0x00 };

        Log.i("HCEDEMO", "Application selected" + new String(apdu));
        //Toast.makeText(MainActivity.this, "selected " + new String(apdu), Toast.LENGTH_SHORT).show();
        Toast.makeText(MainActivity.this, "NXP HCE Test demo succeed", Toast.LENGTH_SHORT).show();
        sendResponseApdu(responseApdu);
        return responseApdu;
    }
    @Override
    public void onDeactivated(int reason) {

    }
}
```

资源文件 apduservice.xml 如下：

```xml
<host-apdu-service xmlns:android="http://schemas.android.com/apk/res/android"
    android:description="@string/servicedesc"
    android:requireDeviceUnlock="false"
    android:apduServiceBanner="@drawable/ic_launcher" >

    <aid-group android:category="payment" android:description="@string/aiddescription" >
        <aid-filter android:name="F0010203040506" />
        <aid-filter android:name="F0394148148100" />
    </aid-group>

    <aid-group android:category="other" >
        <!-- PSE/PPSE AID -->
        <aid-filter android:name="315041592E5359532E4444463031" android:de-
```

```xml
scription = "@string/PSE"/>
            <aid-filter android:name = "325041592E5359532E4444463031" android:de-
scription = "@string/PPSE"/>

            <!-- China Telecom Bestpay AID -->
            <aid-filter android:name = "D1560000401000" android:description = "@
string/ChinaTelecom_Yipay"/>

            <!-- China UnionPay Debit/Credit/Quasi Credit/Electronic Cash AID -->
            <aid-filter android:name = "A000000333010101" android:description = "@
string/UnionPay_Debit"/>
            <aid-filter android:name = "A000000333010102" android:description = "@
string/UnionPay_Credit"/>
            <aid-filter android:name = "A000000333010103" android:description = "@
string/UnionPay_Quasi_Credit"/>
            <aid-filter android:name = "A000000333010106" android:description = "@
string/UnionPay_Electronic_Cash"/>

            <!-- China Mobile AND Wallet Transit/ AID -->
            <aid-filter android:name = "9156000014010001" android:description = "@
string/CMCC_AND_Wallet_Transit"/>

            <!-- China Residents health card/WeiSheng application -->
            <aid-filter android:name = "57532E5359532E4444463031" android:descrip-
tion = "@string/RHC_WeiSheng"/>

            <!-- China Social Insurance -->
            <aid-filter android:name = "7378312E73682EC9E7BBE1B1A3D5CF" android:de-
scription = "@string/Social_Insurance"/>

        </aid-group>

    </host-apdu-service>
```

资源文件 nxp_nfc_ext.xml 如下：

```xml
<extensions xmlns:android = "http://www.nxp.com"
            android:description = "@string/servicedesc">
```

```xml
<nxp-se-ext-group>
    <se-id name="UICC"/>
    <se-power-state name="SwitchOn" value="true"/>
    <se-power-state name="SwitchOff" value="true"/>
    <se-power-state name="BatteryOff" value="false"/>
</nxp-se-ext-group>
</extensions>
```

资源文件 AndroidManifest.xml 如下:

```xml
<?xml version="1.0" encoding="utf-8"?>
<manifest xmlns:android="http://schemas.android.com/apk/res/android"
    package="com.example.hce"
    android:versionCode="1"
    android:versionName="0.2">

    <uses-sdk
        android:minSdkVersion="19"
        android:targetSdkVersion="19"/>

    <uses-permission android:name="android.permission.NFC"/>

<!--    <uses-feature     -->
<!--    android:name="android.hardware.nfc.hce"    -->
<!--    android:required="true"/>    -->

    <application
        android:allowBackup="true"
        android:icon="@drawable/ic_launcher"
        android:label="@string/app_name"
        android:theme="@style/AppTheme">

        <service
            android:name=".MainActivity"
            android:exported="true"
            android:permission="android.permission.BIND_NFC_SERVICE">
            <intent-filter>
                <action android:name="android.nfc.cardemulation.action.HOST_APDU_SERVICE"/>
```

```
            < category android:name = "android.intent.category.DEFAULT" / >
        < /intent - filter >

            < meta - data android:name = "android.nfc.cardemulation.host_apdu_service"
android:resource = "@xml/apduservice" / >
            < meta - data android:name = "com.nxp.nfc.extensions" android:resource = "@
xml/nxp_nfc_ext"/ >

        < /service >

    < /application >

< /manifest >
```

（5）程序端嵌入式方式

这种方式属于一种在线程序内支付方式，例如在一个网页中已经嵌入了访问 SE 单元的证书，当用户刷新网页时出现支付按钮，这时网页调用之前与 SE 中支付程序签过名的 SDK，SDK 通过调用网络接口再转换到 SmartcardService 上，然后再一层层走到下面与 SE 安全单元进行数据交换，完成程序内的支付应用场景。

5.2 TEE 与 eSE 的应用支付流程示例

目前，在国内的 NFC 移动支付方面主要有两大类的应用：一类为符合 PBOC 标准的应用，主要应用在银行以及少数城市的公交系统；另一类为类似选文件的交易模式，有些地域的应用可能还会在非接触的 ISO14443 - 3 或者 ISO14443 - 4 参数上面做一些限制，主要用于各地公交系统。

公交应用区别于银行方面的地方有交易的频率高、金额小等特点，所以在每次刷公交卡时都要去确认指纹，确实不是一个好的用户体验，银行应用需要用户进行指纹确定交易，而公交应用刷卡交易即可。基于这些目标，曾尝试在交易过程中通过 POS 端选择 AID 模式或选择 FID 的模式来区分银行卡和公交卡的应用。但是经过确认和验证，在许多 POS 机上如果交易已经到了选择 AID 或 FID 的环节，如果不马上去完成剩下的交易，那么 POS 机会提示出错。这里无法插入指纹上的区分环节，其原因为在有些地方的公交系统本身就支持双系统，对 PBOC 是同时支持的，所

5 基于 Android 系统的 NFC 实现框架

以通过这种模式做区分走不通。当用户同时下载和安装了公交和银行应用时，银行系统在做非接触工作时，一上来就会选择 PPSE，所以银行与公交应用可以同时为激活状态；但是一般不推荐同时激活的做法，因为在外部的支付环境中可能会存在一些复合交易的场景，支付结束后推荐对该卡片进行激活处理。

下面还是以恩智浦公司的 PN80T 产品为例，如表 5-6 所列为一种 NFC 支付产品的定义。它是一种在现在不区分是银行还是公交应用的产品，在交易时一律要求提供指纹确认，与传统的 Android 的使用习惯的区别在于关机时不支持刷卡，但是要求同时支持 eSE 和 UICC 的卡模拟功能，交易行为习惯如下。

表 5-6 支付行为定义

手机状态	指纹需求	CE@eSE[①]	CE@UICC[①]	CE@HCE[①]
关机	×	×	×	×
开机锁屏状态	√	√	√	×
开机解锁状态	√	√	√	√
开机灭屏状态	√	√	√	×

① 可工作的条件为已经通过了指纹的验证，表中的第三列到最后一列分别为，以 eSE 为卡模拟载体、以 UICC 为卡模拟载体和以 HCE 为卡模拟载体。

手机在靠近 POS 机后支付时会提示用户一个指纹确定的过程，指纹确定无误后 POS 机再和手机完成交易，具体的交易过程流程图如图 5-7 所示，同时适用于银行和公交应用。

交易过程详解如下：

（1）初始以及默认状态部分

①[1] eSE 的电源处于关闭状态，并且没有激活的支付应用在前台。

①[2] NFC 控制器此刻的 CE 是关闭的，包括 UICC 和 eSE 的卡模拟。

（2）提示指纹验证部分

① 在手机靠近 POS 机后，在 NFCC 端能够感应到射频场的开场信号，并且会给 AP 端 REE 侧一条 RF_FIELD_INFO_NTF（Field On 61070101）/NFC_RF_FIELD_REVT(0x5007) 的事件通知。

② AP 端 REE 侧的 NFC 系统应用在拿到这个 Field On 的事件后，会在屏幕提示用户进行指纹确定，TEE 处于激活状态，等待并捕获来自 FP 输入的指纹数据。

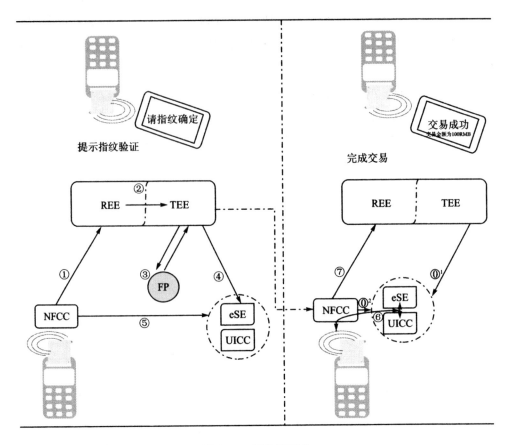

图 5-7 交易流程图

③ FP 接收完整的用户指纹模板输入后,并且运算出指纹特征值送往 TEE FP_TA 进行比对,TEE 侧的 FP_TA 会到 eMMC 中之前存储的 SFS 区域调取出之前存储的指纹特征模板,与刚刚输入的特征值做比对,成功则进入步骤④,反之恢复到 $①^1$、$②^2$ 初始状态。

④ TEE 侧会拉高 B3(SVDD_EN)的电平,给 eSE 通电,并且激活 CRS 相应的 Applet 和开启 eSE 的 CE 模式。

> 如果 SE 上有多个应用,那么在手机的客户端需要支持用户更改默认卡的选项,在设置 CRS 时需要把默认卡选项的参数带进来。

⑤ REE 端发送 NCI 命令开启 NFCC 的 CE 功能,命令如下:

关闭 CE RF_DISCOVER_CMD (2103050283018501)
开启 CE RF_DISCOVER_CMD (2103050280018101)

5 基于 Android 系统的 NFC 实现框架

（3）完成交易

⑥ POS 机再次发起询卡命令后,手机端会有相应的 ISO14443 和应用的响应,POS 机与 eSE 建立安全通道后,完成交易。

⑦ 交易完成后,NFCC 会给 AP 端的 REE 侧一条 EVT_TRANSACTION 的交易事件,并且会把成功标识和成功交易的金额通知给手机系统应用。

⑩^1、⑩^2 交易成功后或者手机离场后,TEE 会把 B3(SVDD_EN)的电平拉低,并去激活非接触模式且 CRS 会把之前激活的应用做激活,通过 REE 去关闭 NFCC 的 UICC 和 eSE 的卡模拟功能。

（4）纠错机制

- 在上面的交易过程中,开启 eSE 电源和 Activation Applet 只有一个条件,就是指纹验证通过才可以,并且一旦通过指纹验证后,在 AP 侧会开启一个定时器,如果在 1 分钟内没有发生交易,则系统也会进入到⑩^1、⑩^2 初始状态,比如指纹验证完了,而用户把手机拿走了,并没有去做交易。
- 若交易步骤中有系统级的错误,比如 REE 收到外部 POS 机交易失败的信号,则这时需要系统进入到⑩^1、⑩^2 初始状态,防止外部恶意攻击。

表 5-7 所列为分解之后的代码步骤。

表 5-7 代码步骤

步骤	功能
1	开关 NFC 所有 CE 卡模拟功能的接口
2	NFC 射频前端控制器当接收到外部场强信号时,上传 Field On 的事件给 REE
3	REE 侧当收到 Field On 后进行弹窗提示用户输入指纹,并把 Wallet 中设置的默认卡的 AID 传递给 TEE 的 TA 程序
4	TEE 侧开启等待和捕获 FP 的匹配,并且在 FP 匹配成功后进行下一步动作
5	TEE 侧会拉高 NFC 射频前端控制器 B3(SVDD_EN)的电平逻辑,给 eSE 通电,并且拿刚刚由 REE 传递过来的默认卡的 AID 去激活 CRS 相应的 Applet
6	TEE 侧要开启一个定时器设置,用于在超时后不管什么原因引起的没有去和 POS 机做交易,TEE 侧都要恢复 CE 卡模拟功能到初始状态
7	NFC 射频前端控制器会给主机端的 REE 侧一条 EVT_TRANSACTION 的交易事件,并且会把成功标识和成功交易的金额通知给手机系统应用,上层用户界面显示交易状态
8	恢复 CE 到初始状态

开关卡模拟的代码示例参数如下：

private static final int EnableCe = 0x00140000;

private static final int EnableLNTCe = 0x00040000;

//LNT to disable type Bprivate static final int DisableCe = 0x0C000000;

/* Combined NFC Technology and protocol bit mask */

#define NFA_DM_DISC_MASK_PA_T1T	0x00000001
#define NFA_DM_DISC_MASK_PA_T2T	0x00000002
#define NFA_DM_DISC_MASK_PA_ISO_DEP	0x00000004
#define NFA_DM_DISC_MASK_PA_NFC_DEP	0x00000008
#define NFA_DM_DISC_MASK_PB_ISO_DEP	0x00000010
#define NFA_DM_DISC_MASK_PF_T3T	0x00000020
#define NFA_DM_DISC_MASK_PF_NFC_DEP	0x00000040
#define NFA_DM_DISC_MASK_P_ISO15693	0x00000100
#define NFA_DM_DISC_MASK_P_B_PRIME	0x00000200
#define NFA_DM_DISC_MASK_P_KOVIO	0x00000400
#define NFA_DM_DISC_MASK_PAA_NFC_DEP	0x00000800
#define NFA_DM_DISC_MASK_PFA_NFC_DEP	0x00001000
#define NFA_DM_DISC_MASK_P_LEGACY	0x00002000

/* Legacy/proprietary/non-NFC Forum protocol (e.g Shanghai transit card) */

| #define NFA_DM_DISC_MASK_POLL | 0x0000FFFF |

#define NFA_DM_DISC_MASK_LA_T1T	0x00010000
#define NFA_DM_DISC_MASK_LA_T2T	0x00020000
#define NFA_DM_DISC_MASK_LA_ISO_DEP	0x00040000
#define NFA_DM_DISC_MASK_LA_NFC_DEP	0x00080000
#define NFA_DM_DISC_MASK_LB_ISO_DEP	0x00100000
#define NFA_DM_DISC_MASK_LF_T3T	0x00200000
#define NFA_DM_DISC_MASK_LF_NFC_DEP	0x00400000
#define NFA_DM_DISC_MASK_L_ISO15693	0x01000000
#define NFA_DM_DISC_MASK_L_B_PRIME	0x02000000
#define NFA_DM_DISC_MASK_LAA_NFC_DEP	0x04000000
#define NFA_DM_DISC_MASK_LFA_NFC_DEP	0x08000000
#define NFA_DM_DISC_MASK_L_LEGACY	0x10000000
#define NFA_DM_DISC_MASK_LISTEN	0xFFFF0000
#define NFA_DM_DISC_MASK_NFC_DEP	0x0C481848

//快速激活卡片的NCI命令过程示例，下面X为主机端发送到NFC射频前端控制器的数据，

//R为NFC射频前端控制器上送到主机端的数据。

5 基于 Android 系统的 NFC 实现框架

X: 4 > 21060100 RF_DEACTIVATE_CMD

R: 4 > 41060100 RF_DEACTIVATE_RSP

X: 8 > 2103050283018501

[NCI][COMMAND][21 03 05 02 83 01 85 01]

RF_DISCOVER_CMD

* Number of Configurations = 2 [0x02]

-- > Configuration N° 1

- RF Technology and Mode = NFC_A_ACTIVE_LISTEN_MODE[0x83]

- Discovery Frequency = RF Technology and Mode will be executed in every discovery period [0x01]

-- > Configuration N° 2

- RF Technology and Mode = NFC_F_ACTIVE_LISTEN_MODE[0x85]

- Discovery Frequency = RF Technology and Mode will be executed in every discovery period [0x01]

R: 4 > 41030100 RF_DISCOVER_RSP

X: 5 > 220102C000 NFCEE_MODE_SET_CMD

R: 4 > 42010100 NFCEE_MODE_SET_RSP

X: 5 > 220102C001 NFCEE_MODE_SET_CMD

R: 4 > 42010100 NFCEE_MODE_SET_RSP

APDU DATA_PACKET/CORE_CONN_CREDITS_NTF 开关逻辑通道

X:10 > 03000799500070000001 //open logical channel

R: 6 > 600603010301

R: 8 > 0300059950019000 //open logical channel success

X:12 > 030009995080CA00FE02DF21 //GET SERIAL NUMBER 软件上电,通过 SPI 写入 SET CRS

R: 6 > 600603010301

R:36 > 0300219950FE1BDF211804324073213880015357004518937374 5A8AFB2E54826D599000

X: 9 > 030006995001708001//close logical channel

R: 6 > 600603010301

R: 7 > 03000499509000//close logical channel success

X: 5 > 0300029961

R: 6 > 600603010301

APDU DATA_PACKET/CORE_CONN_CREDITS_NTF 开关逻辑通道

X: 4 > 21060100 RF_DEACTIVATE_CMD

R: 4 > 41060100 RF_DEACTIVATE_RSP

X: 8 > 2103050280018101

[NCI][COMMAND][21 03 05 02 80 01 81 01]

RF_DISCOVER_CMD

* Number of Configurations = 2 [0x02]

-- > Configuration N° 1

- RF Technology and Mode = NFC_A_PASSIVE_LISTEN_MODE[0x80]

- Discovery Frequency = RF Technology and Mode will be executed in every discovery period [0x01]

-- > Configuration N° 2

- RF Technology and Mode = NFC_B_PASSIVE_LISTEN_MODE[0x81]

- Discovery Frequency = RF Technology and Mode will be executed in every discovery period [0x01]

R: 4 > 41030100 RF_DISCOVER_RSP

X: 4 > 21060100 RF_DEACTIVATE_CMD

R: 4 > 41060100 RF_DEACTIVATE_RSP

X: 4 > 2F150102 (DISPLAY_OFF_STATE_CMD) = Screen Locked

R: 4 > 4F150100

X: 8 > 2103050280018101

[NCI][COMMAND][21 03 05 02 80 01 81 01]

RF_DISCOVER_CMD

* Number of Configurations = 2 [0x02]

-- > Configuration N° 1

- RF Technology and Mode = NFC_A_PASSIVE_LISTEN_MODE[0x80]

- Discovery Frequency = RF Technology and Mode will be executed in every discovery period [0x01]

-- > Configuration N° 2

- RF Technology and Mode = NFC_B_PASSIVE_LISTEN_MODE[0x81]

- Discovery Frequency = RF Technology and Mode will be executed in every discovery period [0x01]

R: 4 > 41030100 RF_DISCOVER_RSP

X: 4 > 21060100 RF_DEACTIVATE_CMD

R: 4 > 41060100 RF_DEACTIVATE_RSP

X: 8 > 2103050280018101

[NCI][COMMAND][21 03 05 02 80 01 81 01]

RF_DISCOVER_CMD

5 基于 Android 系统的 NFC 实现框架

```
* Number of Configurations = 2 [0x02]
-- > Configuration N°  1
- RF Technology and Mode = NFC_A_PASSIVE_LISTEN_MODE[0x80]
- Discovery Frequency = RF Technology and Mode will be executed in every discovery
  period [0x01]

-- > Configuration N°  2
- RF Technology and Mode = NFC_B_PASSIVE_LISTEN_MODE[0x81]
- Discovery Frequency = RF Technology and Mode will be executed in every discovery
  period [0x01]
R: 4 > 41030100 RF_DISCOVER_RSP
```

5.3 卡模拟应用程序示例

前面章节介绍过卡模拟基于的载体也是各式各样的,本节主要为基于恩智浦公司的 P73 的 JCOP 平台进行开发的示例。图 5-8 是需要在开发集成环境中嵌入的 SDK 包和 API 接口等,这里只是一个示例,具体引用的数据包需要查看具体的产品型号和 JCOP 的版本。

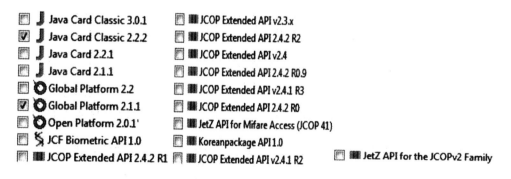

图 5-8　JCOP SDK 插件包

如图 5-9 所示,在创建一个测试支付的私有的 Package 和 Applet 时,需要指定的 AID 完全符合 ISO7816-5 要求,其由应用注册标识(5 字节)＋私有应用扩展标识(0～11 字节)组成,不要少于 5 字节。

下面示例了一个当得到一个外部 APDU 数据命令时,函数 process 开始响应处理,并且调用 SE 安全单元提供的 JAVA 接口进行的 AES 加解密。

Package:
com.xiaomi.mipay

Package AID:

Applet AID:

图 5-9　Applet 和 Package 包指定示意

命令脚本文件示例如下：

```
/select |aesApp
/send 80200000 *9000
```

源文件 AESAppletExample.java 如下：

```java
package com.xiaohua.test.aes;


import javacard.framework.APDU;
import javacard.framework.Applet;
import javacard.framework.ISO7816;
import javacard.framework.ISOException;
import javacard.security.AESKey;
import javacard.security.KeyBuilder;
import javacard.security.Signature;
import javacardx.crypto.Cipher;


public class AESAppletExample extends Applet {
    final byte[] key1 = new byte[] {(byte)0x10,(byte)0x00,(byte)0x00,(byte)0x00,
(byte)0x00,(byte)0x00,(byte)0x00,(byte)0x00,(byte)0x00,(byte)0x00,(byte)0x00,(byte)
0x00,(byte)0x00,(byte)0x00,(byte)0x00,(byte)0x00};
    final byte[] key2 = new byte[] {(byte)0x20,(byte)0x00,(byte)0x00,(byte)0x00,
(byte)0x00,(byte)0x00,(byte)0x00,(byte)0x00,(byte)0x00,(byte)0x00,(byte)0x00,(byte)
0x00,(byte)0x00,(byte)0x00,(byte)0x00,(byte)0x00};
    final byte[] key3 = new byte[] {(byte)0x30,(byte)0x00,(byte)0x00,(byte)0x00,
(byte)0x00,(byte)0x00,(byte)0x00,(byte)0x00,(byte)0x00,(byte)0x00,(byte)0x00,(byte)
0x00,(byte)0x00,(byte)0x00,(byte)0x00,(byte)0x00};
```

```
        final byte[] key4 = new byte[] {(byte)0x40,(byte)0x00,(byte)0x00,(byte)0x00,
(byte)0x00,(byte)0x00,(byte)0x00,(byte)0x00,(byte)0x00,(byte)0x00,(byte)
0x00,(byte)0x00,(byte)0x00,(byte)0x00,(byte)0x00};

        AESKey[] aesKey = new AESKey[4];
        Cipher EncCipher, DecCipher;

        public AESAppletExample(){
            // get a reference to an AESKey
            for(byte i = 0; i < 4; i++)
            aesKey[i] = (AESKey) KeyBuilder.buildKey(KeyBuilder.TYPE_AES, KeyBuilder.
LENGTH_AES_128, false);
            aesKey[0].setKey(key1,(short)0);
            aesKey[1].setKey(key2,(short)0);
            aesKey[2].setKey(key3,(short)0);
            aesKey[3].setKey(key4,(short)0);

            // get a reference to a Cipher which will do AES encryption/decryption
            EncCipher = Cipher.getInstance(Cipher.ALG_AES_BLOCK_128_CBC_NOPAD, false);
            DecCipher = Cipher.getInstance(Cipher.ALG_AES_BLOCK_128_CBC_NOPAD, false);
        }

        public static void install(byte[] bArray, short bOffset, byte bLength) {
            // GP - compliant Java Card applet registration
            new AESAppletExample().register(bArray, (short) (bOffset + 1), bArray[bOff-
set]);
        }

        public void process(APDU apdu) {

            short myP1;
            // Good practice: Return 9000 on SELECT
            if (selectingApplet()) {
                return;
            }

            // reference to the APDU buffer
            byte[] buf = apdu.getBuffer();
```

```
// check CLA byte
if(buf[ISO7816.OFFSET_CLA] ! = (byte) 0x80)
    ISOException.throwIt(ISO7816.SW_CLA_NOT_SUPPORTED);


// check INS byte
switch (buf[ISO7816.OFFSET_INS]) {
// INTIALIZE AES key and Cipher
case (byte) 0x10:
    // receive data field
    apdu.setIncomingAndReceive();
    // populate AES key
    aesKey[0].setKey(buf, ISO7816.OFFSET_CDATA);
    // initialize Cipher with encryption
    EncCipher.init(aesKey[0], Cipher.MODE_ENCRYPT);
    break;
// Encrypt command data field, must be block aligned (no padding)
case (byte) 0x20:
    // receive data field
    apdu.setIncomingAndReceive();
    myP1 = buf[ISO7816.OFFSET_P1];
    if(myP1 > 3)
        ISOException.throwIt(ISO7816.SW_DATA_INVALID);

    EncCipher.init(aesKey[0], Cipher.MODE_ENCRYPT);
    // encrypt 16 byte data
    EncCipher.doFinal(buf, ISO7816.OFFSET_CDATA, (short)0x10, buf, (short)0);
    // respond with the encrypted data
    apdu.setOutgoingAndSend((short) 0, (short) 0x10);
    break;

    // Decrypt data field,...
case (byte) 0x30:
    apdu.setIncomingAndReceive();
    myP1 = buf[ISO7816.OFFSET_P1];
    if(myP1 > 3)
    ISOException.throwIt(ISO7816.SW_DATA_INVALID);

    DecCipher.init(aesKey[0], Cipher.MODE_DECRYPT);
```

```
            // encrypt 16 byte data
            DecCipher.doFinal(buf, ISO7816.OFFSET_CDATA, (short)0x10, buf, (short)0);
            // respond with the encrypted data
            apdu.setOutgoingAndSend((short) 0, (short) 0x10);
            break;
        default:
            // good practice: If you don't know the INStruction, say so:
            ISOException.throwIt(ISO7816.SW_INS_NOT_SUPPORTED);
        }
    }
}

/**
 *
 * 1. Create a second Cipher to decrypt, extend instructions, and verify encryption
 * 2. Create an object array and populate it with 4 different AES keys
 * 3. Let the reader choose what key will be used, P1 parameter
 * 4. Create and test a Signature based on AES
 *
 */
```

安全通道 SCP02 应用程序实现代码示例。

命令脚本文件示例如下：

```
/echo ${time}
/select |NXP_Test|01
/echo "Authentication"
init-upd
ext-auth crmac
/echo "Read 1"
send 80CA9E000400010101
/echo "Read 2"
send 80CA9E000C0602080107010903030202O1
/echo "Read 3"
send 80CA9E00060B08040D0C05
/echo "Write 1"
send
80E20000A99E00A70A06230A0A0A0A0A0A0A0A0A0A0A0A0A0A0A0A0A0A0A0A0A0A0A0A0A0A0A0A0A0A0A0
```

A0A0A0A0A0A0A0A0B04230B0B0B0B0B0B0B0B0B0B0B0B0B0B0B0B0B0B0B0B0B0B0B0B0B0
B0B0B0B0B0B0B080125080808080808080808080808080808080808080808080808080808080
8080808080808080806020C0606060606060606060606000120FAFAFAFAFAFAFAFAFAFAFAFA
FAFAFAFAFAFAFAFAFAFAFAFAFAFAFAFA00
/echo "check Write 1"
send 80CA9E000A0A060B04080106020001

源文件 SCP02DemoApplet.java 如下：

```java
package com.nxp.bli.jcop.javacard.cas.scp02;

import javacard.framework.APDU;
import javacard.framework.Applet;
import javacard.framework.ISO7816;
import javacard.framework.ISOException;
import javacard.framework.JCSystem;
import javacard.framework.Util;
import javacard.security.DESKey;
import javacard.security.KeyBuilder;
import javacard.security.RandomData;
import javacard.security.Signature;
import javacardx.crypto.Cipher;

public class SCP02DemoApplet extends Applet {

    private static final byte INS_INITUPDATE = (byte)0x50;
    private static final byte INS_EXTAUTH    = (byte)0x82;

    private static final short OFFSET_CMAC_FLAG = (short)0;
    private static final byte CMAC_RECALCULATION_REQUIRED = (byte)0xA5;
    private static final byte CMAC_RECALCULATION_NOT_NEEDED = (byte)0x5A;
    private static final short OFFSET_SELECT_FLAG = (short)1;
    private static final byte SELECT_APDU_BEFORE = (byte)0x00;
    private static final byte OTHER_APDU_BEFORE = (byte)0x81;
    private static final short OFFSET_SECURE_CHANNEL_FLAG = (short)2;
    private static final byte SECURE_CHANNEL_OPENED = (byte)0xCA;
    private static final byte SECURE_CHANNEL_CLOSED = (byte)0xFE;
    private static final short OFFSET_INIT_UPD_FLAG = (short)3;
    private static final byte INIT_UPD_BEFORE = (byte)0xCC;
```

```java
//      private static final short OFFSET_RMAC_FLAG = (short)4;
//      private static final byte RMAC_RECALCULATION_REQUIRED = (byte)0x4A;
//      private static final byte RMAC_RECALCULATION_NOT_NEEDED = (byte)0xA4;
        private static final short OFFSET_SEC_LEVEL = (short)5;
        private static final byte NO_SECURITY = (byte)0x00;
        private static final byte C_MAC_ONLY = (byte)0x01;
        private static final byte C_AND_R_MAC = (byte)0x11;

        private static final byte SECURE_CHANNEL_PROTOCOL = (byte)0x02;

        private static final short LENGTH_OF_HOST_CHALLENGE = (short)8;
        private static final short LENGTH_OF_APDU_HEADER = (short)4;
        private static final short LENGTH_OF_MAC = (short)8;
        private static final short LENGTH_OF_ICV = (short)8;
//      private static final short LENGTH_OF_FINAL_BLOCK = (short)8;
        private static final short LENGTH_OF_HOSTCRYPTOGRAM_AND_MAC = (short)((short)8 + LENGTH_OF_MAC);

        private static final byte[] DES_PADDING = new byte[]{(byte)0x80, (byte)0x00, (byte)0x00, (byte)0x00, (byte)0x00, (byte)0x00, (byte)0x00, (byte)0x00};

        private static final byte[] DERIVATION_DATA_ENC_KEY = new byte[]{(byte)0x01, (byte)0x82};
        //private static final byte[] DERIVATION_DATA_DENC_KEY = new byte[]{(byte)0x01, (byte)0x81};
        private static final byte[] DERIVATION_DATA_CMAC_KEY = new byte[]{(byte)0x01, (byte)0x01};
        private static final byte[] DERIVATION_DATA_RMAC_KEY = new byte[]{(byte)0x01, (byte)0x02};

        private byte[] s_ENC_key;
        private byte[] s_MAC_key;
        //private byte[] s_DEK_key;

        private byte keyVersionNumber;

        DESKey encKey;
        DESKey cmacKey;
```

```java
        DESKey rmacKey;
        DESKey tmpKey;
        DESKey icvKey;
        Cipher tmpCipher;
        Signature cmacSignature;

        private short sequenceCounter; // sequence counter shall be from 0 to 65535
        private byte[] cardChallenge;
        private byte[] keyDiversicationData;
        private byte[] lastCMAC;
        private byte[] lastRMAC;
        private byte[] flagBuf;
        private byte[] icvKeyData;
        private byte[] tempCmacBuf;

        // EF and records are hard coded
        private byte[] EF_ALL;

    // Tag
        // Single tags
        private static final byte EF_FAT_TAG = (byte) 0x0;
        private static final byte EF_CARD_HEADER_TAG = (byte) 0x1;
        private static final byte EF_ENVIRONMENT_TAG = (byte) 0x2;
        private static final byte EF_HOLDER_TAG = (byte) 0x3;
        private static final byte EF_CONTRACT_FIX_TAG = (byte) 0x4;
        private static final byte EF_LOYALTYLIST_TAG = (byte) 0x5;
        private static final byte CtxAndPurse_TAG = (byte) 0x6;
        private static final byte EF_CONTRACTLIST_TAG = (byte) 0x7;
        private static final byte EF_SPECIALEVENTLIST_TAG = (byte) 0x8;
        private static final byte EF_TPURSELOG_TAG = (byte) 0x9;
        private static final byte EF_EVENTLOG_TAG = (byte) 0xA;
        private static final byte EF_SPECIALEVENT_TAG = (byte) 0xB;
        private static final byte EF_CONTRACT_VAR_TAG = (byte) 0xC;
        private static final byte EF_LOYALTY_TAG = (short) 0xD;

        private static final byte MAX_TAG = EF_LOYALTY_TAG;

        // Number of entries
```

5 基于 Android 系统的 NFC 实现框架

```java
    private static final byte[] entryArray = {
        1, 1, 2, 2, 15, 2, 2, 2, 2, 3, 11, 14, 15, 9
    };
// length of entries
    private static final byte[] lengthArray = {
        32, 20, 13, 24, 48, 11, 12, 32, 37, 35, 35, 35, 13, 16
    };

    private static short[] offsetArray;
    private static short EF_ALL_LENGTH;

    // DGI for multiple read/write of single tags
    // CLA INS < P1P2 = 0x9E00 > Lc Tag1 EntryNo Tag2 EntryNo .. -- > even length
    private static final short DGI_MULT_SINGLE = (short) 0x9E00;

    // CLA
    private static final byte CLA_AUTH = (byte)0x80;
    private static final byte CLA_MAC_ENC = (byte)0x84;

    // INS
    private static final byte INS_GET_DATA = (byte)0xCA; // read
    private static final byte INS_STORE_DATA = (byte)0xE2; // write

    // transient byte array
    private byte[] swapBuf;

    public static void install(byte[] bArray, short bOffset, byte bLength) {
        new com.nxp.bli.jcop.javacard.cas.scp02.SCP02DemoApplet().register(bArray,
(short)(bOffset + 1), bArray[bOffset]);
    }

    /**
     * Create the needed cipher objects for all further cipher operations
     * @throws Exception
     */
    public SCP02DemoApplet(){
        // set key values to default
        this.keyVersionNumber = (byte)0xFF;
```

```
                this.s_ENC_key = new byte[]{(byte)0x40,(byte)0x41,(byte)0x42,(byte)0x43,
(byte)0x44,(byte)0x45,(byte)0x46,(byte)0x47,
                                            (byte)0x48,(byte)0x49,(byte)0x4A,(byte)
0x4B,(byte)0x4C,(byte)0x4D,(byte)0x4E,(byte)0x4F};
                this.s_MAC_key = new byte[]{(byte)0x40,(byte)0x41,(byte)0x42,(byte)0x43,
(byte)0x44,(byte)0x45,(byte)0x46,(byte)0x47,
                                            (byte)0x48,(byte)0x49,(byte)0x4A,(byte)
0x4B,(byte)0x4C,(byte)0x4D,(byte)0x4E,(byte)0x4F};
                //this.s_DEK_key = new byte[]{(byte)0x40,(byte)0x41,(byte)0x42,(byte)
0x43,(byte)0x44,(byte)0x45,(byte)0x46,(byte)0x47,
                //                            (byte)0x48,(byte)0x49,(byte)0x4A,(byte)
0x4B,(byte)0x4C,(byte)0x4D,(byte)0x4E,(byte)0x4F};
                // get key objects to 'store' session keys
                this.encKey = (DESKey)KeyBuilder.buildKey(KeyBuilder.TYPE_DES_TRANSIENT_DE-
SELECT,KeyBuilder.LENGTH_DES3_2KEY,false);

                this.cmacKey = (DESKey)KeyBuilder.buildKey(KeyBuilder.TYPE_DES_TRANSIENT_
DESELECT,KeyBuilder.LENGTH_DES3_2KEY,false);

                this.rmacKey = (DESKey)KeyBuilder.buildKey(KeyBuilder.TYPE_DES_TRANSIENT_
DESELECT,KeyBuilder.LENGTH_DES3_2KEY,false);

                this.icvKey = (DESKey)KeyBuilder.buildKey(KeyBuilder.TYPE_DES_TRANSIENT_DE-
SELECT,KeyBuilder.LENGTH_DES,false);

                this.tmpKey = (DESKey)KeyBuilder.buildKey(KeyBuilder.TYPE_DES_TRANSIENT_DE-
SELECT,KeyBuilder.LENGTH_DES3_2KEY,false);

                // card challenge, incl. padding
                this.cardChallenge = new byte[]{(byte)0x11,(byte)0x22,(byte)0x33,(byte)
0x44,(byte)0x55,(byte)0x66,(byte)0x80,(byte)0x00};

                this.keyDiversicationData = new byte[]{(byte)0x01,(byte)0x02,(byte)0x03,
(byte)0x04,(byte)0x05,(byte)0x06,(byte)0x07,(byte)0x08,(byte)0x09,(byte)0x0A};

                this.lastCMAC = JCSystem.makeTransientByteArray((short)8,JCSystem.CLEAR_ON_
DESELECT);
                this.lastRMAC = JCSystem.makeTransientByteArray((short)8,JCSystem.CLEAR_ON_
DESELECT);
                this.flagBuf = JCSystem.makeTransientByteArray((short)7,JCSystem.CLEAR_ON_
DESELECT);
                this.icvKeyData = JCSystem.makeTransientByteArray((short)8,JCSystem.CLEAR_
```

```
ON_DESELECT);
            // also used for RMAC
            this.tempCmacBuf = JCSystem.makeTransientByteArray((short)16,JCSystem.
CLEAR_ON_DESELECT);
            // swapBuf for GET DATA and storage for R_MAC calculation
            // -- > 260 byte C_APDU + length R_APDU data field + + R_APDU data field + 2
byte SW1SW2 + padding = 520 bytes max
            swapBuf = JCSystem.makeTransientByteArray((short)520,JCSystem.CLEAR_ON_
DESELECT);

            this.tmpCipher = Cipher.getInstance(Cipher.ALG_DES_CBC_NOPAD,false);
            // also used for RMAC
            this.cmacSignature = Signature.getInstance(Signature.ALG_DES_MAC8_ISO9797_
1_M2_ALG3,false);

            offsetArray = new short[entryArray.length];
            short swap = 0;
            short i = 1;
            do {
                swap = offsetArray[i-1];
                offsetArray[i] = (short)(swap + lengthArray[i-1] * entryArray[i-1]);
                i++;
            } while(i < offsetArray.length);
            EF_ALL_LENGTH = (short)(offsetArray[i-1] + lengthArray[i-1] * entryArray
[i-1]);

            EF_ALL = new byte[EF_ALL_LENGTH]; // 2349 byte
            RandomData rnd = RandomData.getInstance(RandomData.ALG_PSEUDO_RANDOM);
            rnd.generateData(EF_ALL,(short)0,(short)EF_ALL.length);

            for(i = 0,swap = 0; i < entryArray.length; i++)
                swap += entryArray[i];
    }

    public void process(APDU apdu) {
        short swap = 0; // < -- our fast local variable!
        short offset = 0;
        short dataLength = apdu.setIncomingAndReceive();
```

```java
            if (selectingApplet()) {
                this.flagBuf[OFFSET_CMAC_FLAG] = CMAC_RECALCULATION_NOT_NEEDED;
                this.flagBuf[OFFSET_SECURE_CHANNEL_FLAG] = SECURE_CHANNEL_CLOSED;
                RandomData rnd = RandomData.getInstance(RandomData.ALG_PSEUDO_RANDOM);
                // only for test purposes, fill data with random numbers <-- takes long time
                rnd.generateData(EF_ALL, (short)0, (short) EF_ALL.length);
                return;
            }

            byte[] buf = apdu.getBuffer();
            switch (buf[ISO7816.OFFSET_INS]) {
            case INS_INITUPDATE:
                this.flagBuf[OFFSET_CMAC_FLAG] = CMAC_RECALCULATION_NOT_NEEDED;
                this.flagBuf[OFFSET_SECURE_CHANNEL_FLAG] = SECURE_CHANNEL_CLOSED;
                // clear CLEAR_ON_DESELECT data if no SELECT before
                if(this.flagBuf[OFFSET_SELECT_FLAG] != SELECT_APDU_BEFORE){
                    Util.arrayFillNonAtomic(this.lastCMAC, (short)0, (short)this.lastCMAC.length, (byte)0);
                    Util.arrayFillNonAtomic(this.lastRMAC, (short)0, (short)this.lastCMAC.length, (byte)0);
                    Util.arrayFillNonAtomic(this.icvKeyData, (short)0, (short)this.icvKeyData.length, (byte)0);
                }
                this.flagBuf[OFFSET_SELECT_FLAG] = OTHER_APDU_BEFORE;
                // extract host challenge
                if(dataLength != LENGTH_OF_HOST_CHALLENGE){
                    ISOException.throwIt(ISO7816.SW_WRONG_LENGTH);
                }
                // create session keys
                // APDU buffer construction
                // |0      ...    13|14    ...    29|30   ...    45|
                // | incoming APDU |derivation data|tmp key value|
                //----------------------------------------------
                // ENC-Key
                Util.arrayCopyNonAtomic(DERIVATION_DATA_ENC_KEY, (short)0, buf, (short)14, (short)2);
                Util.setShort(buf, (short)16, this.sequenceCounter);
```

```java
this.tmpKey.setKey(this.s_ENC_key,(short)0);
this.tmpCipher.init(this.tmpKey,Cipher.MODE_ENCRYPT);
this.tmpCipher.doFinal(buf,(short)14,(short)16,buf,(short)30);
Util.arrayCopyNonAtomic(buf,(short)30,buf,(short)46,(short)8);
this.encKey.setKey(buf,(short)30);
// CMAC - Key
Util.arrayCopyNonAtomic(DERIVATION_DATA_CMAC_KEY,(short)0,buf,(short)14,(short)2);
Util.setShort(buf,(short)16,this.sequenceCounter);
this.tmpKey.setKey(this.s_MAC_key,(short)0);
this.tmpCipher.init(this.tmpKey,Cipher.MODE_ENCRYPT);
this.tmpCipher.doFinal(buf,(short)14,(short)16,buf,(short)30);
this.cmacKey.setKey(buf,(short)30);
//------------------------------------------
// RMAC - Key
Util.arrayCopyNonAtomic(buf,(short)30,this.icvKeyData,(short)0,(short)8);
Util.arrayCopyNonAtomic(DERIVATION_DATA_RMAC_KEY,(short)0,buf,(short)14,(short)2);
Util.setShort(buf,(short)16,this.sequenceCounter);
this.tmpKey.setKey(this.s_MAC_key,(short)0);
this.tmpCipher.init(this.tmpKey,Cipher.MODE_ENCRYPT);
this.tmpCipher.doFinal(buf,(short)14,(short)16,buf,(short)30);
this.rmacKey.setKey(buf,(short)30);
//------------------------------------------
// create card cryptogram
// APDU buffer construction
// |30   ...   37|38   ...   39|40   ...   45|46   ...   53|  < -- buf
// |Host Challenge|Sequence Counter|Card Challenge|DES padding|
Util.arrayCopyNonAtomic(buf,(short)5,buf,(short)30,(short)8);
Util.setShort(buf,(short)38,this.sequenceCounter);
Util.arrayCopyNonAtomic(this.cardChallenge,(short)0,buf,(short)40,(short)6);
Util.arrayCopyNonAtomic(DES_PADDING,(short)0,buf,(short)46,(short)8);
// we can save time if we reuse the initialized cipher
// for the encryption of the card challenge used for verification in EXT - AUTH
// |0   ...   1|2   ...   8|9   ...   16|  < -- swapBuf
```

```java
            // |Sequence Counter|Card Challenge|Host Challenge|
            Util.arrayCopyNonAtomic(buf,(short)38,swapBuf,(short)0,(short)8);
// counter + card challenge
            Util.arrayCopyNonAtomic(buf,(short)30,swapBuf,(short)8,(short)8);
// host challenge
            Util.arrayCopyNonAtomic(DES_PADDING,(short)0,swapBuf,(short)16,
(short)8); // DES Padding
            this.tmpCipher.init(this.encKey,Cipher.MODE_ENCRYPT);
            this.tmpCipher.doFinal(buf,(short)30,(short)24,buf,(short)4); //
card cryptogram
            this.tmpCipher.doFinal(swapBuf,(short)0,(short)24,swapBuf,(short)
0); // host cryptogram
            // build response APDU
            // APDU buffer construction
            // |0 ... 9|  10  |11|12 ... 13|14 ... 19|20 ... 27|
            // |Key Diversification Data|Key Version Number|SCP|Sequence Counter|
Card Challenge|Card Cryptogram|
            Util.arrayCopyNonAtomic(this.keyDiversicationData,(short)0,buf,
(short)0,(short)10);
                buf[10] = this.keyVersionNumber;
                buf[11] = SECURE_CHANNEL_PROTOCOL;
                Util.setShort(buf,(short)12,this.sequenceCounter);
                Util.arrayCopyNonAtomic(this.cardChallenge,(short)0,buf,(short)14,
(short)6);
                apdu.setOutgoingAndSend((short)0,(short)28);
                this.flagBuf[OFFSET_INIT_UPD_FLAG] = INIT_UPD_BEFORE;
                break;
        case INS_EXTAUTH:
            this.flagBuf[OFFSET_SECURE_CHANNEL_FLAG] = SECURE_CHANNEL_CLOSED;
            if(this.flagBuf[OFFSET_INIT_UPD_FLAG] != INIT_UPD_BEFORE){
                this.flagBuf[OFFSET_CMAC_FLAG] = CMAC_RECALCULATION_NOT_NEEDED;
                ISOException.throwIt(ISO7816.SW_CONDITIONS_NOT_SATISFIED);
            }
            this.flagBuf[OFFSET_INIT_UPD_FLAG] = OTHER_APDU_BEFORE;
            // CLA BYTE = GP + SM
            if(buf[ISO7816.OFFSET_CLA] != (byte)0x84){
                this.flagBuf[OFFSET_CMAC_FLAG] = CMAC_RECALCULATION_NOT_NEEDED;
                ISOException.throwIt(ISO7816.SW_CLA_NOT_SUPPORTED);
```

```
            }
            // Minimum C_MAC (or CRMAC?)
            if((buf[ISO7816.OFFSET_P1] & (byte)1) != C_MAC_ONLY) {
                this.flagBuf[OFFSET_CMAC_FLAG] = CMAC_RECALCULATION_NOT_NEEDED;
                ISOException.throwIt(ISO7816.SW_SECURITY_STATUS_NOT_SATISFIED);
            } else if(buf[ISO7816.OFFSET_P1] == C_MAC_ONLY)
                this.flagBuf[OFFSET_SEC_LEVEL] = C_MAC_ONLY;
            else if(buf[ISO7816.OFFSET_P1] == C_AND_R_MAC)
                this.flagBuf[OFFSET_SEC_LEVEL] = C_AND_R_MAC;
            else {
                this.flagBuf[OFFSET_CMAC_FLAG] = CMAC_RECALCULATION_NOT_NEEDED;
                ISOException.throwIt(ISO7816.SW_SECURITY_STATUS_NOT_SATISFIED);
            }
            if(dataLength != LENGTH_OF_HOSTCRYPTOGRAM_AND_MAC){
                this.flagBuf[OFFSET_CMAC_FLAG] = CMAC_RECALCULATION_NOT_NEEDED;
                ISOException.throwIt(ISO7816.SW_WRONG_LENGTH);
            }
            if(! this.encKey.isInitialized()){
                this.flagBuf[OFFSET_CMAC_FLAG] = CMAC_RECALCULATION_NOT_NEEDED;
                ISOException.throwIt(ISO7816.SW_SECURITY_STATUS_NOT_SATISFIED);
            }
            // 1st verify host cryptogram (encrypted card challenge)
            // swapBuf holds our encrypted challenge, prepared in INIT-UPD
            if(Util.arrayCompare(buf, ISO7816.OFFSET_CDATA, swapBuf, (short)16,
(short)8) != 0){
                this.flagBuf[OFFSET_SECURE_CHANNEL_FLAG] = SECURE_CHANNEL_CLOSED;
                ISOException.throwIt(ISO7816.SW_SECURITY_STATUS_NOT_SATISFIED);
            }
            // 2nd verify CMAC on EXT-AUTH
            this.verifyCMAC(buf, dataLength);
            // increment sequence counter to avoid 'Replay Attack'!
            if(sequenceCounter != (short)0xFFFF)
                this.sequenceCounter++;
            // this MAC will be the initial ICV for R-MAC
            Util.arrayCopyNonAtomic(lastCMAC, (short)0, lastRMAC, (short)0, LENGTH_OF_MAC);
            this.flagBuf[OFFSET_SECURE_CHANNEL_FLAG] = SECURE_CHANNEL_OPENED;
            break;
```

```
case INS_GET_DATA: // read
    this.flagBuf[OFFSET_INIT_UPD_FLAG] = OTHER_APDU_BEFORE;
    if(this.flagBuf[OFFSET_SECURE_CHANNEL_FLAG] != SECURE_CHANNEL_OPENED){
        ISOException.throwIt(ISO7816.SW_CONDITIONS_NOT_SATISFIED);
    }
    if(dataLength < LENGTH_OF_MAC){
        this.flagBuf[OFFSET_CMAC_FLAG] = CMAC_RECALCULATION_NOT_NEEDED;
        ISOException.throwIt(ISO7816.SW_WRONG_LENGTH);
    }
    // need a second buffer since the APDU buffer data field will be overwritten, but is needed for R_MAC calculation
    Util.arrayCopyNonAtomic(buf, (short)0, swapBuf, (short)0, dataLength);
    this.verifyCMAC(buf, dataLength); // apdu buffer is trashed now
    // remove secure messaging bit and logical channel in CLA byte, C_MAC and adjust Lc <-- basically GP's unwrap method
    swapBuf[ISO7816.OFFSET_CLA] &= (byte)0xF8;
    swapBuf[ISO7816.OFFSET_LC] = (byte)((dataLength - 8) & 0x00FF);
    dataLength -= (short)8;
    if(Util.makeShort(swapBuf[ISO7816.OFFSET_P1], swapBuf[ISO7816.OFFSET_P2]) == DGI_MULT_SINGLE) {
        if((dataLength & (short)1) != 0) // no data field or data field length is uneven
            ISOException.throwIt(ISO7816.SW_CONDITIONS_NOT_SATISFIED);
        Util.setShort(buf, (short)0, DGI_MULT_SINGLE);
        swap = 3;
        for(short i = ISO7816.OFFSET_CDATA; i < ISO7816.OFFSET_CDATA + dataLength; i += 2) {
            // tag value and record number exists?
            if((offset = getTagOffset(swapBuf[i], swapBuf[i + 1])) < (short)0)
                ISOException.throwIt(ISO7816.SW_FILE_NOT_FOUND);
            buf[swap] = swapBuf[i]; // tag
            swap++;
            buf[swap] = swapBuf[i + 1]; // record
            swap++;
            byte length = lengthArray[swapBuf[i]];
            buf[swap] = length; // tag length
            swap++;
            Util.arrayCopyNonAtomic(EF_ALL, offset, buf, swap, length);
```

```
                    // value
                            swap += length;
                    }
                    buf[2] = (byte)(swap - 2);
            } else { // just one tag, encoded in P1 and record number in P2
                    if((swap = getTagOffset(swapBuf[ISO7816.OFFSET_P1], swapBuf
[ISO7816.OFFSET_P2])) < (short)0)
                            ISOException.throwIt(ISO7816.SW_FILE_NOT_FOUND);
                    buf[0] = buf[ISO7816.OFFSET_P1]; // tag
                    buf[1] = lengthArray[buf[ISO7816.OFFSET_P1]]; // length
                    Util.arrayCopyNonAtomic(EF_ALL, swap, buf, (short)2, lengthArray
[buf[ISO7816.OFFSET_P1]]); // value
                    swap = (short)(buf[1] + 2);
            }
            if(this.flagBuf[OFFSET_CMAC_FLAG] != C_AND_R_MAC){
                    swap = generateRmac(buf, swapBuf, swap, (short)(dataLength +
ISO7816.OFFSET_CDATA));
            }

            apdu.setOutgoingAndSend((short)0, swap);
            break;
        case INS_STORE_DATA: // write
            this.flagBuf[OFFSET_INIT_UPD_FLAG] = OTHER_APDU_BEFORE;
            if(this.flagBuf[OFFSET_SECURE_CHANNEL_FLAG] != SECURE_CHANNEL_OPENED){
                    ISOException.throwIt(ISO7816.SW_CONDITIONS_NOT_SATISFIED);
            }
            if(dataLength < LENGTH_OF_MAC){
                    this.flagBuf[OFFSET_CMAC_FLAG] = CMAC_RECALCULATION_NOT_NEEDED;
                    ISOException.throwIt(ISO7816.SW_WRONG_LENGTH);
            }
            // need a second buffer since the APDU buffer data field will be overwrit-
ten, but is needed for R_MAC calculation
            Util.arrayCopyNonAtomic(buf, (short)0, swapBuf, (short)0, dataLength);
            this.verifyCMAC(buf, dataLength); // apdu buffer is trashed now
            // remove secure messaging bit and logical channel in CLA byte, C_MAC and
adjust Lc < -- basically GP's unwrap method
            swapBuf[ISO7816.OFFSET_CLA] &= (byte)0xF8;
            swapBuf[ISO7816.OFFSET_LC] = (byte)((dataLength - 8) & 0x00FF);
```

```
                    dataLength -= (short)8;
                if(dataLength > 6 + ISO7816.OFFSET_CDATA && Util.makeShort(buf[ISO7816.
OFFSET_CDATA], buf[ISO7816.OFFSET_CDATA + (byte)1]) == DGI_MULT_SINGLE) {
                    // check DGI length field
                    if(!(buf[ISO7816.OFFSET_CDATA + (byte)2] == (byte)(dataLength - 2)))
                        ISOException.throwIt(ISO7816.SW_WRONG_LENGTH);
                    for(swap = (ISO7816.OFFSET_CDATA + 3); swap < dataLength;) {
                        // APDU buffer, on the position 0: tag, 1: record, 2: length, 3: value
                        // tag value and record number exists?
                        if((offset = getTagOffset(buf[swap], buf[swap + 1])) < (short)0)
                            ISOException.throwIt(ISO7816.SW_FILE_NOT_FOUND);
                        // check length field
                        if(buf[swap + 2] != lengthArray[buf[swap]])
                            ISOException.throwIt(ISO7816.SW_WRONG_LENGTH);
                        // is the data available?
                        if(dataLength + ISO7816.OFFSET_CDATA <= (swap + buf[swap + 2] + 2))
                            ISOException.throwIt(ISO7816.SW_WRONG_LENGTH);
                        if(buf[swap] == EF_FAT_TAG) {
                            // write atomic for FAT (tear protected)
                            Util.arrayCopy(buf, (short)(swap + 3), EF_ALL, offset,
lengthArray[buf[swap]]);
                        } else { // not tear protected
                            Util.arrayCopyNonAtomic(buf, (short)(swap + 3), EF_ALL,
offset, lengthArray[buf[swap]]);
                        }
                        swap = (short)(swap + buf[swap + 2] + 3);
                    }
                } else {
                    if(dataLength < 3 && (offset = getTagOffset(buf[ISO7816.OFFSET_
CDATA], buf[ISO7816.OFFSET_CDATA + (byte)1])) < 0)
                        ISOException.throwIt(ISO7816.SW_FILE_NOT_FOUND);
                    // check length field
                    if(buf[ISO7816.OFFSET_CDATA + 2] != lengthArray[buf[ISO7816.OFFSET_
CDATA]])
                        ISOException.throwIt(ISO7816.SW_WRONG_LENGTH);
                    // is the data available?
                    if(dataLength < buf[ISO7816.OFFSET_CDATA + 2])
                        ISOException.throwIt(ISO7816.SW_WRONG_LENGTH);
```

5 基于 Android 系统的 NFC 实现框架

```java
                    // ready to write, atomic
                    Util.arrayCopy(buf, (short)(ISO7816.OFFSET_CDATA + 3), EF_ALL, off-
set, lengthArray[buf[ISO7816.OFFSET_CDATA]]);
                }
                if(this.flagBuf[OFFSET_CMAC_FLAG] != C_AND_R_MAC){
                    swap = generateRmac(buf, swapBuf, (short)0, (short)(dataLength +
ISO7816.OFFSET_CDATA));
                    apdu.setOutgoingAndSend((short)0, LENGTH_OF_MAC);
                }
                break;
            default:
                this.flagBuf[OFFSET_CMAC_FLAG] = CMAC_RECALCULATION_NOT_NEEDED;
                this.flagBuf[OFFSET_INIT_UPD_FLAG] = OTHER_APDU_BEFORE;
                ISOException.throwIt(ISO7816.SW_INS_NOT_SUPPORTED);
        }
    }


    private void verifyCMAC(byte[] buf, short lc){
        if(lc < LENGTH_OF_MAC){
            return;
        }
        short posDataEnd = (short)(LENGTH_OF_APDU_HEADER + (short)1 + lc - (short)8);
        // move MAC, needed for final comparison: 0..7 = CMAC - 8..15 = temp. ICV
        Util.arrayCopyNonAtomic(buf, posDataEnd, tempCmacBuf, (short)0, LENGTH_OF_MAC);

        if(this.flagBuf[OFFSET_CMAC_FLAG] == CMAC_RECALCULATION_NOT_NEEDED){
            Util.arrayCopyNonAtomic(this.lastCMAC, (short)0, tempCmacBuf, LENGTH_OF_
MAC, LENGTH_OF_ICV);
        }else{
            // GP requires ICV encryption for C_MAC
            this.icvKey.setKey(this.icvKeyData, (short)0);
            this.tmpCipher.init(this.icvKey, Cipher.MODE_ENCRYPT);
            this.tmpCipher.doFinal(this.lastCMAC, (short)0, LENGTH_OF_MAC, tempC-
macBuf, LENGTH_OF_ICV);
        }
        // GP 2.1.1 C-MAC
        cmacSignature.init(cmacKey, Signature.MODE_SIGN, tempCmacBuf, LENGTH_OF_
MAC, LENGTH_OF_ICV);
```

```java
            cmacSignature.sign(buf, (short) 0, posDataEnd, buf, posDataEnd);
            if(Util.arrayCompare(buf, (short)(posDataEnd), tempCmacBuf, (short) 0,
LENGTH_OF_MAC) != (byte)0){
                this.flagBuf[OFFSET_CMAC_FLAG] = CMAC_RECALCULATION_NOT_NEEDED;
                this.flagBuf[OFFSET_SECURE_CHANNEL_FLAG] = SECURE_CHANNEL_CLOSED;
                this.flagBuf[OFFSET_SEC_LEVEL] = NO_SECURITY;
                ISOException.throwIt(ISO7816.SW_SECURITY_STATUS_NOT_SATISFIED);
            }
            this.flagBuf[OFFSET_CMAC_FLAG] = CMAC_RECALCULATION_REQUIRED;
            // last CMAC = next ICV
            Util.arrayCopyNonAtomic(tempCmacBuf, (short) 0, this.lastCMAC, (short) 0,
LENGTH_OF_MAC);
        }


        private short generateRmac(byte[] responseBuf, byte[] commandBuf, short response-
Length, short commandLength) {
            // Li
            commandBuf[commandLength] = (byte)(responseLength & 0x00FF);
            // SW12 also signed
            responseBuf[responseLength] = (byte)0x90;
            responseBuf[responseLength + 1] = (byte)0x00;
            // GP2.1.1 R-MAC
            cmacSignature.init(rmacKey, Signature.MODE_SIGN, lastRMAC, (short)0, LENGTH_
OF_ICV);
            // first the command
            cmacSignature.update(commandBuf, (short) 0, (short)(commandLength + 1));
            // then response with SW12
            cmacSignature.sign(responseBuf, (short) 0, (short)(responseLength + 2),
responseBuf, responseLength);
            // last RMAC = next ICV
            Util.arrayCopyNonAtomic(responseBuf, responseLength, this.lastRMAC, (short)
0, LENGTH_OF_MAC);
            return (short)(responseLength + 8);
        }

        private short getTagOffset(byte tag, byte entry) {
            if(tag < 0 || tag > MAX_TAG)
                return -1;
```

```
            if(entryArray[tag] < entry)
                return -2;
            else
                return ((short)(offsetArray[tag]+((entry-1) * lengthArray[tag])));
    }
}
```

6　附　录

本书内容参考了以下资料：

- ISO International Organization for Standardization

  http://www.iso.org/iso/home.html

- IEC International Electrotechnical Commission

  http://www.iec.ch/

- ETSI/TS European Telecommunications Standards Institute Technical Specification

  http://www.etsi.org/WebSite/homepage.aspx

- Ecma European Computer Manufacturers Association TC47

  http://www.ecma-international.org/memento/TC47-M.htm

- NFC Forum

  - Committees andworking groups

    http://www.nfc-forum.org/aboutus/committees/

  - Specifications list

    http://www.nfc-forum.org/specs/spec_dashboard/

- GP Global Platform

  https://www.globalplatform.org/home.asp

- Java API Specifications

  https://www.oracle.com/java/java-card.html

- China UnionPay

  http://cn.unionpay.com/

- EMVCo

  http://www.emvco.com/specifications.aspx

- OSCCA

  http://www.oscca.gov.cn/

- PCI DSS

https://www.pcisecuritystandards.org/documents/

- NXP semiconductors

    http://www.nxp.com/

- STMicroelectronics

    http://www.st.com/content/st_com/en.html

- Broadcom Limited

    https://www.broadcom.com/

- Oberthur Technologies

    http://www.oberthur.com

- Android SDK for windows

    http://developer.android.com/sdk/index.html

- Android open source

    http://source.android.com/

- Eclipse SDK

    http://www.eclipse.org/downloads/

- Java SE downloads

    http://www.oracle.com/technetwork/java/javase/downloads/index.html

- Ubuntu 11.04 (Natty Narwhal)

    http://ie.releases.ubuntu.com/natty/

- Python

    http://www.python.org/download/

- Virtualbox

    http://www.virtualbox.org/wiki/Downloads

- Git

    http://git-scm.com/download

- Valgrind

    http://valgrind.org/downloads/current.html

- Seek

    http://code.google.com/p/seek-for-android/

- android.nfc

http://developer.android.com/reference/android/nfc/package-summary.html
- android.nfc.tech

  http://developer.android.com/reference/android/nfc/tech/package-summary.html
- 《NFC 技术基础篇》

  https://item.jd.com/12108117.html
- 《Paying With Plastic: The Digital Revolution in Buying and Borrowing》

  《塑料卡片的魔力:交易的数字化革命》
- MST

  http://www.looppay.com/

  http://www.samsung.com
- Prochip

  http://www.prochip.com.cn/
- MagChip

  http://www.idtechproducts.com/
- Singular

  http://www.singular.com.tw/
- Magtek

  https://www.magtek.com
- GB/T 33736—2017

  《手机支付 基于 2.45 GHz RCC(限域通信)技术的非接触射频接口技术要求》
- JR/T 0025.4—2012《中国金融集成电路(IC)卡规范》
- CJ/T 304—2008《建设事业 CPU 卡操作系统技术要求》
- 《交通一卡通移动支付技术规范》
- 《城市公共交通 IC 卡技术规范》
- GM/T 0001—2012《祖冲之序列密码算法》
- GM/T 0002—2012《SM4 分组密码算法》(原 SMS4 分组密码算法)
- GM/T 0003—2012《SM2 椭圆曲线公钥密码算法》
- GM/T 0004—2012《SM3 密码杂凑算法》
- GM/T 0005—2012《随机性检测规范》
- GM/T 0006—2012《密码应用标识规范》